Going Vegan
To Save The Planet

Going Vegan
To Save The Planet

By M.E Whitehead

© 2025 Montoya Whitehead

All rights reserved.

No portion of this book may be reproduced, copied, distributed or adapted in any way, with the exception of certain activities permitted by applicable copyright laws, such as brief quotations in the context of a review. For permission to publish, distribute or otherwise reproduce this work, please contact the author at montoyawhitehead@gmail.com

This publication is designed to provide accurate and authoritative information regarding the subject matter covered. It is sold with the understanding that neither the author nor the publisher is engaged in rendering legal, investment, accounting or other professional services. While the publisher and author have used their best efforts in preparing this book, they make no representations or warranties with respect to the accuracy or completeness of the contents of this book and specifically disclaim any implied warranties of merchantability or fitness for a particular purpose. No warranty may be created or extended by sales representatives or written sales materials. The advice and strategies contained herein may not be suitable for your situation. You should consult with a professional when appropriate. Neither the publisher nor the author shall be liable for any loss of profit or any other commercial damages, including but not limited to special, incidental, consequential, personal, or other damages.

ISBN 978-1-9191940-0-4

Cover by: Montoya Whitehead

Illustrations by: Francisco Atencio

Dedication

To my nan, Sheila.

If she hadn't taken that first step, I wouldn't be doing this.

Table of Contents

Introduction _____ *9*

Chapter 1: Understanding Climate Change: History, Causes and Future _____ *17*

*Chapter 2: Why We Now Say Climate Change Instead of Global Warming*_____ *38*

Chapter 3: Reducing Water Pollution Through Veganism _____ *52*

Chapter 4: Pollution, Poultry Farming and the Planet _____ *88*

*Chapter 5: Fishing and Farming the Oceans*_____ *124*

Chapter 6: Burning Forests for Beef _____ *164*

Chapter 7: Hogs and Harms _____ *204*

Chapter 8: Dairy Farming's Dirty Footprint _____ *252*

*Chapter 9: Wool, Feathers, Skins and Fur*_____ *298*

Chapter 10: It's Not Just Animal Agriculture That's The Problem *354*

Chapter 11: Putting it All Together _____ *371*

References _____ *390*

Introduction

A Planet in Peril

The Earth is at its most critical juncture. Human activity has driven our planet almost to the point of no return; we are at a cliff edge. If we don't make changes, we will fall off the cliff edge into destruction. For centuries, humans have used the planet's resources to fuel growth, innovation and prosperity. But this progress has come at a steep cost. Today, we are witnessing the devastating consequences of our actions: rising temperatures, rising sea levels, extreme weather events, deforestation, loss of biodiversity and widespread pollution. These issues, once just concerns, are now real threats to the ecosystems that support life on Earth.

In March 1924, the average sea surface temperature for the planet was ⁻0.26°C; by February 2024, this temperature had risen to 0.95°C. To take a more narrow view, in Italy January 1940, the average surface air temperature, 2 metres above ground, was 0.75°C. In January 2024, the average surface temperature was 6.12°C.

Agriculture is the largest producer of methane (CH_4), a potent greenhouse gas. In 2020, agriculture alone produced 3.54 billion tonnes of methane. Agriculture, by far, is the largest producer of nitrous oxide in the form of animal waste. In 2020, farmers produced 2.33 billion tonnes of nitrous oxide, using it as a fertiliser in the form of manure. This nitrous oxide goes on to pollute soil and water, with the potential to harm human health. In 2022, the planet's temperature had risen by 1.64°C since records began. Agriculture and land use cause 0.58°C of that global temperature increase. Put another way, agriculture and land use are responsible for 0.58°C, more than 35%, of global warming. The planet is in peril, and we are the danger. We are simultaneously heating and poisoning the Earth, with no real or effective plan to stop.

Don't think that this won't affect you. This isn't a "them" problem; this isn't "I'm ok, I like hot summers". Climate change impacts us all. It's not just an environmental issue; it's a complex crisis that touches every part of our lives. It impacts the food we eat, the water we drink, the air we breathe, and even the stability of our societies. The science is clear. If we do not drastically reduce our carbon emissions and change our environmental practices, we risk reaching a point of no return. A scenario where the damage becomes irreversible and the survival of future generations is at grave risk. We've already lost countless animal species; they are gone forever. People have lost their lives. Entire communities have been displaced by sea level rise. Forests and countless animals are being destroyed by wildfires. Yet we continue as if everything is fine.

The Power of Choice

The thing about climate change is that we can all do something to help, to mitigate the damage, if not turn it around, and there is a growing recognition that individual choices do matter. While government policies, technological advancements and global cooperation are crucial, the decisions we make in our everyday lives can also have a profound impact on the environment. One of the most significant areas where personal choices intersect with environmental outcomes is our diet.

The food we consume is not only a source of nourishment; it is also a key driver of environmental change. The production and processing of food has a significant impact on the planet's resources. Agriculture, particularly animal farming, is a leading contributor to greenhouse gas emissions, deforestation and pollution. Animal agriculture uses a significant share of global water and causes land degradation and biodiversity loss.

In this context, veganism - a lifestyle that avoids the use of animal products - emerges as a powerful tool for environmental sustainability. By choosing a plant-based diet, we can reduce our ecological footprint, conserve resources and contribute to the fight against climate change. Veganism is not just a diet; it is a broader commitment to reducing harm to animals and the environment, encompassing everything from food to clothing.

What Is Veganism

The way we choose to eat has a profound impact on the environment. It also affects our health and how we treat animals. There are many dietary lifestyles that people can choose from today, including veganism, vegetarianism, pescatarianism and omnivorous diets. These lifestyles represent distinct approaches to sourcing nutrition and sustaining life. They reflect differing priorities, such as minimising harm to animals, optimising health, or balancing cost with convenience.

Omnivorism

An omnivorous diet includes plant-based foods, fruits, vegetables, grains, etc. and animal-based foods, such as meat, dairy and eggs. Eating in this way offers flexibility, with pretty much all foods being on the table. While it can be balanced and healthy, diets high in meat come with nitrates and fats. Nitrates, saturated fats and cholesterol are all linked to chronic diseases. There is also the concern with processed and ultra-processed foods, although these are not only animal-based products. Ethical and environmental concerns also arise with industrial meat production, which contributes to deforestation, greenhouse gas emissions and animal welfare issues. The impact that an omnivore's diet can have depends largely on the balance between plant and animal foods and the sustainability of the farming methods.

Pescatarianism

Pescatarianism is a way of eating that incorporates seafood along with plant-based foods, sometimes including dairy and eggs. Many see it as a healthier way to eat and a compromise between vegetarianism and omnivorism. They aim to reduce the reliance on land animals while benefiting from the omega-3 fatty acids found in fish. Proponents claim that pescatarian diets support heart and brain health. However, there are concerns about overfishing and mercury levels in certain fish that can make this diet less healthy, more expensive, and less sustainable.

Vegetarianism

Vegetarianism excludes meat and fish, but still includes animal by-products such as dairy and eggs, though variations exist. Lacto-ovo vegetarians consume both dairy and eggs, while lacto-vegetarians avoid eggs but still consume dairy and dairy products. Ovo vegetarians exclude all dairy while still consuming eggs. Many people choose vegetarianism for ethical reasons, believing that no animals are harmed, while also believing that humans need animal proteins to be healthy. Some people focus on the diet's health benefits, including lower risks of heart disease and hypertension, even though every diet, from an omnivore to a vegan diet, has the potential to be unhealthy.

Veganism

Veganism is a dietary and lifestyle choice that excludes all animal products, including meat, fish, dairy, and eggs as well as by-products such as honey. Beyond food, vegans often avoid animal-derived items such as leather and wool. This reflects a commitment to improved animal welfare and an end to animal exploitation. While ethical vegans focus on ending animal exploitation, environmental vegans focus on the diet's reduced ecological footprint. Health-conscious vegans often adopt this diet, favouring whole foods, such as brown rice over white rice, to lower the risks of chronic diseases. Vegan diets have an emphasis on plant-based proteins, fruits, vegetables, grains and nuts. But supplementation for nutrients such as vitamin B12 is necessary. This is due to modern living eliminating this from traditional sources, such as water used for drinking.

Why This Book?

This book is a response to the growing need for clear, accessible information on how our dietary and lifestyle choices affect the environment. It connects the complicated science of climate change with simple actions people can take. Some books discuss veganism for health. They focus on lowering cholesterol and reducing the risk of heart attacks or strokes. There are many books on veganism about the harm done to animals and how animal agriculture is bad for the environment. Some books challenge the arguments against veganism. They answer questions about canine teeth and whether humans should have control over animals. Few books directly discuss how going vegan is one of the most powerful ways to slow or even reverse climate change.

The book, *Food and Climate Change without the Hot Air: Change your Diet: The Easiest Way to Help Save the Planet,* by S.L. Bridle, examines the greenhouse gas emissions from different meals. This includes items like a slice of buttered toast and a bowl of cereal. The ingredients and cooking methods are investigated to identify their environmental impact. The book then gives alternatives that result in fewer emissions. Books like *The Green Path: A Practical Guide to Sustainable Living* by K. Connors advise how to reduce your environmental footprint. They suggest ways to minimise waste and maintain a more sustainable lifestyle. As great as these books are, and I have no criticisms, they do not delve into the impacts of animal agriculture. They do not address how a vegan lifestyle, avoiding all animal products, including wool and leather, is good for our health and the planet.

Imagine a world where rainforests thrive. Where record-high temperatures are written about in history books. Regions, laid to waste by droughts, become lush, and there's enough food for everyone. This doesn't have to be a fantasy; we can do this. With consistent, small, individual acts, we can reduce greenhouse gases. Choosing to make a vegan chilli, chilli no carne (chilli sin carne really), or having a couple of vegan sausages instead of pork with your breakfast are simple. But they are also impactful, reducing your water, land and carbon footprint. This book is for you, whether you're vegan, thinking about going vegan, or just curious about sustainable living. It offers knowledge and inspiration to help you take action.

We will explore the environmental impacts of the different industries connected to animal farming. We'll cover the meat, dairy, poultry, fishing and fashion industries. Each chapter explores how these industries harm the environment and drive climate change. We will see how this is supported by scientific evidence and case studies with expert insights. You will learn how these industries operate, the scale of their impact, and how adopting a vegan lifestyle can mitigate these effects.

The Urgency of Now

The urgency of addressing climate change cannot be overstated, as those living in affected areas already know. The Intergovernmental Panel on Climate Change (IPCC) has issued stark warnings. We must limit global warming to 1.5°C above pre-industrial levels. This requires unprecedented changes across all sectors of society, including agriculture and food production. Switching to a plant-based diet is one of the best ways to cut greenhouse gas emissions and help the environment.

But beyond the statistics and scientific models, this book is about hope and empowerment. It is about recognising that, despite the enormity of the challenge, each of us has the power to make a positive impact. We need to rethink our relationship with food and make more sustainable choices. We can contribute to a healthier planet for ourselves and future generations. Simple changes, choosing a chickpea curry instead of chicken, cooking lentils in place of minced beef, baking with flaxseed egg substitute (just flaxseed and water) rather than chicken eggs, are things that we can do. They are often more nutritious and lower in saturated fats. But they are also better for the planet, requiring less land and water to produce while having a lower carbon footprint.

With something as important as the future of our existence at stake and a solution that is so easy and attainable, it is only apathy that stands in our way. Grandparents dying with their grandchild is far from apathetic. Their house collapsed beneath them as they waited on the roof for rescue from severe flooding caused by Hurricane Helene. This hurricane was one of the deadliest to hit the United States. Its strength has been linked directly to climate change. There is nothing apathetic about wildfires destroying entire communities. There is nothing apathetic about people freezing to death in their homes in Texas during a snowstorm. There is no place for apathy in an emergency.

A Journey Towards Sustainability

As you start your journey through this book, you will see how all life on Earth is connected. You will understand how the food on your plate is linked to the health of our planet. The clothes you wear can affect ecosystems. Your choices as a consumer can create positive change.

The transition to a more sustainable way of living does not have to be overwhelming. Small, incremental changes can lead to significant impacts. There are many books, YouTube channels and social media accounts that can help you through the process. This book is not about blaming or vilifying any individual. There are large corporations and entire industries that profit from the public not knowing how agriculture is damaging the planet. With this knowledge, we can make informed choices about our lives and the impact we create.

A Call to Action

This book is both a source of information and a call to action. It challenges you to think critically about your role in the environmental crisis. It encourages you to think about how going vegan can greatly help the environment. The choices we make today will shape the world of tomorrow. By choosing compassion, sustainability and responsibility, we can build a future where people and the planet thrive.

As you turn the pages, keep an open mind and heart. Let the ideas and information here inspire you to act. Together, we can make a difference. No matter if you're thinking about veganism, just starting, or deepening your commitment to sustainable living, remember that every choice counts and every action matters. We can be the change.

Chapter 1: Understanding Climate Change: History, Causes and Future

What is Climate Change?

"Climate change, that's just weather, and the weather changes every day." – my father-in-law

Climate change is a huge and complicated concept. It is about far more than the weather that you see through the window. Climate change is the long-term shift in temperature and weather patterns in a specific region. Natural processes and human activities cause this complex phenomenon. Throughout its history, the Earth's climate has undergone significant changes over millions of years. In science, "climate change" refers to the unprecedented warming since the late 1800s due to human activities. Records show that, since the Industrial

Revolution, carbon dioxide (CO_2) levels in the Earth's atmosphere have risen sharply [1]. Deforestation, burning fossil fuels and industrial processes have caused global temperatures to rise. This has caused a sudden shift in the Earth's climate, fuelling the dramatic weather that we see on the news and experience in our daily lives.

The Science Behind Climate Change

At the heart of climate change lies the greenhouse effect - a natural process that warms the Earth's surface. When the Sun's energy reaches the Earth, one of two things will happen: some energy is reflected out to space, while the rest is absorbed. This energy is then re-radiated by **greenhouse gases (GHGs)** in the atmosphere. These greenhouse gases include **carbon dioxide (CO_2), methane (CH_4) and nitrous oxide (N_2O)**. Before human intervention, this process kept the Earth's temperature stable enough to support life. Yet, human activities have significantly increased the concentration of these gases. This enhances the greenhouse effect and causes the Earth's average temperature to rise, as less heat can escape into space.

Over the past century, global temperatures have risen by approximately **1.2°C (2.2°F)**, with most of the warming occurring in the last few decades [2]. This may seem like a small change, but it has profound effects on the climate system. A rise in temperature of 0.5°C changes the temperature of the seas and oceans, as they act as a heat sink for the planet. The heating leads to more water evaporating into the air and the melting of ice in the far North and South of the planet. This will then affect the winds and jet streams that blow around the planet. The changes cause a domino effect, altering weather patterns, ocean currents and ecosystems.

Historical Context: From the Industrial Revolution to Today

To understand the current climate crisis, it's essential to look at how human activity has changed the Earth's atmosphere over time.

The Industrial Revolution began in the late 18th century in England. It started with the development of a mechanical process for producing cloth. This was a turning point in human history. Societies transitioned away from agrarian economies to industrialised markets. An economy grew around the ability to mass-produce goods. This mass production gave rise to factories and the widespread use of fossil fuels like coal, oil and natural gas. These energy sources were efficient. However, they had a major downside: they released large amounts of carbon dioxide into the atmosphere.

The Industrial Revolution completely changed the way societies worked. Machines and other mechanical instruments meant that the power used to produce goods no longer came from humans or animals. Both people and animals need food and rest; they also have limited abilities. The new machines could work for hours nonstop. They did not need payment after the initial outlay necessary to buy them. When a machine broke down, parts were replaced and the machine was repaired. Machines also had no rights and could not get sick; automation was the way forward. For those who could afford them, machines were a guaranteed wealth generator, as goods were cheaper to produce. As factories opened, they created jobs and wealth. Those who had previously worked making products of their own to sell for little profit now had reliable pay. People with jobs could afford to buy the goods that factories produced, creating a cycle of demand that continues today. The energy needed for all this production came in the form of fossil fuels. Steam, produced by burning coal to heat water, oil, and later gas, powered these advancements. This led to the mass extraction and refining of these fossil fuels, which became industries of their own.

Before the Industrial Revolution, the concentration of CO_2 in the atmosphere was around **280 parts per million (ppm)**. Today, that number has increased to over **420 ppm**, the highest level in at least **800,000 years** [3]. This dramatic increase is directly linked to human activities. Burning fossil fuels, deforestation for agriculture, and industrial processes in particular.

Key Milestones in the Understanding of Climate Change

The link between human activities and climate change has been studied for over a century. These are some key landmark moments:

1. **The Discovery of the Greenhouse Effect (1824-1896):** In 1824, French scientist Joseph Fourier suggested certain gases trap heat in the atmosphere [4]. Later, in 1856, American scientist Eunice Foote suggested that carbon dioxide with water vapour could have a warming effect on the Earth [5][6]. The concept was refined by Swedish scientist Svante Arrhenius in 1896. He calculated the potential impact of CO_2 on global temperatures [7]. He also speculated that the warming would benefit people in the future as they "might live under a milder sky and in less barren surroundings."
2. **The Keeling Curve (1958):** Charles David Keeling began measuring CO_2 levels in the atmosphere at the Mauna Loa Observatory in Hawaii. His data showed a steady increase in CO_2 levels, providing the world with its first clear evidence of human-caused climate change.
3. **The Formation of the IPCC (1988):** The Intergovernmental Panel on Climate Change (IPCC) was established by the World Meteorological Organization and the United Nations. The IPCC provides policymakers with regular assessments of climate change science. This includes

its impacts and future risks. They then use their data to propose strategies for mitigation and adaptation.
4. **The Paris Agreement (2015):** The Paris Agreement has been adopted by almost every country in the world. The goal is to keep global warming below 2°C, with a target of 1.5°C. This landmark agreement marked a significant step towards global climate action.

Current Trajectory: The Warming World

Today, the impacts of climate change are being felt globally. The Earth has already warmed by more than 1.2°C since pre-industrial times [8]. The consequences of this warming are becoming increasingly severe and widespread:

- **Rising Sea Levels:** 90% of the increase in the planet's temperature is occurring in the oceans [9]. Melting ice caps and glaciers, combined with the thermal expansion of seawater, are causing sea levels to rise. This is a direct threat to coastal communities. Low-lying areas like Bangladesh, the Maldives and parts of the United States are particularly at risk. In these regions, some populations are already dealing with their homes being submerged or washed away.

- **Extreme Weather Events:** The frequency and intensity of extreme weather events have increased. This includes the number of hurricanes, floods, droughts and wildfires. These events cause significant disruption. Damage to property and infrastructure, and disruption to food and water supplies, are becoming common. Millions of people are being displaced or losing their lives [9].

- **Ocean Acidification:** The oceans have absorbed about 90% of the atmospheric warming that has already occurred [10]. Ocean acidification is harming marine life. This particularly affects organisms with calcium carbonate shells or skeletons. Organisms, such as corals and certain plankton species, suffer harm as the lower pH of the water dissolves the calcium carbonate.

- **Loss of Biodiversity:** Climate change is altering habitats and putting species at risk [9]. Many plants and animals are struggling to adapt to the changing conditions, leading to a decline in biodiversity. Wildfires can destroy terrain, killing many trees, plants, and creatures, with those that survive having nowhere to live. Wildlife can suffer from burns and smoke inhalation. This is clear in sensitive ecosystems, such as coral reefs, rainforests and the Arctic tundra.

The Global Response

In response to these trends, governments, businesses, and individuals are beginning to act. International accords, including the Paris Agreement, target greenhouse gas emissions to limit global warming. Countries are adopting renewable energy sources, such as solar and wind power, to reduce their dependence on fossil fuels [9]. Additionally, there is growing awareness of the need to protect and restore natural ecosystems. These ecosystems play a critical role in sequestering carbon and maintaining biodiversity. Projects, such as planting trees and mangroves, restore habitats and strengthen ecosystems. They also act as carbon sinks, sequestering carbon from the air and storing it inside plants and trees.

But the pace of change is not fast enough to avoid the worst impacts of climate change. The IPCC warns that limiting global warming to 1.5°C will need "rapid, far-reaching and unprecedented changes in all aspects of society." This includes reducing global carbon emissions by as much as half by 2030 and reaching net zero by 2050.

Future Predictions: What Lies Ahead?

The future impacts of climate change will depend on the actions we take today. Scientists use scenarios, known as Representative Concentration Pathways (RCPs), to model potential outcomes. The scenarios are based on different levels of greenhouse gas emissions:

1. **RCP 2.6:** This scenario assumes that global emissions peak around 2020 and then decline rapidly. If we follow this path, we could limit global warming to about 1.5°C by the end of the century, avoiding some of the most severe impacts of climate change.
2. **RCP 4.5:** In this scenario, emissions peak around 2040 and then decline. Global temperatures would likely rise by around 2.4°C, leading to significant but manageable impacts.
3. **RCP 6.0:** Under this scenario, emissions continue to rise until around 2080. At this point, they stabilise, resulting in warming of about 3°C, with severe consequences for ecosystems and human societies.
4. **RCP 8.5:** The worst-case scenario. Emissions continue to rise throughout the century, leading to catastrophic environmental changes. If we follow this path, global temperatures could increase by more than 4°C. This would lead to catastrophic environmental changes. Widespread displacement and severe economic and social disruption would result.

Sources:
- Met Office [11]
- Carbon Brief [12]

Why Every Degree Matters

One of the most important lessons from climate science is that every fraction of a degree matters. The difference between 1.5°C and 2°C of warming is more significant than it seems. It could mean the difference between survival and extinction for many species. This includes the difference between manageable impacts and widespread disasters for human societies.

For example, 1.5°C of warming could mean the death of most coral reefs as the increase in ocean temperature would cause widespread bleaching. Significant ice melt in the Arctic would raise sea levels. This would erode coastlines and could cause an increase in extreme weather events. At 2°C, these impacts become more severe and acute, with higher sea levels, more intense heat waves and greater risks to global food and water security.

"Wine, is kind of the canary in the coal mine for climate change impacts on agriculture because so much of the character of wine is tied to the local climate." - Benjamin Cook, climate scientist at NASA's Goddard Institute for Space Studies. [13]

1°C Higher

A target of a 2°C global increase was set by climate scientists as the highest temperature that the Earth would be able to recover from. William Nordhaus, a Yale economist, first suggested in the 1970s that a 2°C increase was a significant tipping point. He said that human civilisation had never experienced such global conditions at any point in history. Nordhaus's theory was that a 2°C rise in temperatures could cause extreme conditions. It was more than a decade later that James Hansen, a NASA scientist, testified before Congress. He was one of the first scientists to link human greenhouse gas emissions to rising global temperatures [14]. Hansen stated that if carbon emissions were not reduced, climate change could be catastrophic. The change would cause rising sea levels and extreme weather events. Ecosystems would be destroyed, and human populations would be displaced. He explained that once the comfortable areas of the planet would become uninhabitable.

The Paris Agreement 2015 aimed to limit the global temperature increase to 2°C by reducing greenhouse gas emissions. The purpose was to reduce the risk of the worst outcomes of climate change in most of the world [15]. The IPCC now uses 1.5°C as the target in its reports. If all greenhouse gas emissions stopped today, temperatures would still rise by at least 1°C. This would happen because temperature increases trigger a chain reaction. This chain reaction began over 100 years ago and will slow rather than stop [16]. The argument that the Earth has been this hot before, so this process is natural and thus okay, does not work. 6,000 years ago, temperatures were at this point, 1°C higher. If we were to recreate this scenario, ⅓ of the Earth's freshwater would be gone in 80 years. The snow that covers mountains would melt away, causing rivers to dry up. The lack of freshwater would impact agriculture, with less water available for crops and animals [17]. There would be landslides in highlands and mountainous regions, caused by melting permafrost. Permafrost binds the earth; as it melts, the earth becomes mud. These mud and rockslides would change the landscape, burying anything in their way. Melting permafrost would also release stored methane and CO_2, further accelerating global warming.

At an average temperature raised by 1°C, there would be more frequent and intense heatwaves [15]. Warmer oceans would provide more water and energy for hurricanes and cyclones. Their intensity would increase. Hurricane season would start earlier and last longer. Storms would be more powerful, causing billions of dollars of devastation. Shifts in rain patterns could also cause some regions to experience more droughts. Others would face more frequent flooding and, added to this, would be rising sea levels. Polar ice caps and glaciers would continue to melt, but at a faster rate. An increase of 1°C contributes to the thermal expansion of seawater. This results in further melting of ice, coastal flooding, erosion, and higher sea levels. Higher temperatures would negatively impact marine species in seas and oceans. Coral bleaching would lead to more coral reef deaths [17]. As temperatures rise, many species, including birds and fish, would migrate to cooler regions. This would disrupt ecosystems and food chains. Other species, such as those with limited adaptability or mobility, such as limpets or mussels, could face extinction.

More frequent heatwaves would cause increased rates of illness in human populations [17]. Heatstroke, dehydration, and diseases like malaria and dengue would be more common. They would become common in areas of the world that did not experience such health conditions. In agriculture, some crops would thrive in the warmer weather, while others would struggle. Some regions would face reduced yield and unpredictable harvests, impacting food resources, as well as the economy [15].

We have a limit on how much carbon we can put into the atmosphere before we face an existential crisis: 2.9 trillion tonnes. We have already emitted 1.9 trillion; we have 1 trillion left, and at this rate, we will have used those 1 trillion tonnes of carbon in 21 years.

2°C Higher

At a global temperature increase of 2°C, people start dying [17]. In the summer of 2023, global temperatures were 2.3°C hotter than average. In Europe, 52,000 people died as a result, with the very young and very old being most affected. If temperatures were 2°C hotter, countries like the Maldives would be submerged. Sea levels would rise, and countries, already hit by hurricanes, would face ever greater storms, such as Hurricane Helene in 2024 [16]. Some people, usually not those with any scientific background, argue that plants will grow more. They believe that we will have greater crop yields as CO_2 increases. But at 2°C warmer, plant growth slows, then stops, as plants suffer from heat stress [18]. Heat stress happens when plants experience temperatures that exceed a certain limit for too long. This can cause lasting harm to their growth and development [19]. As animal species migrate to cooler areas, natural patterns of breeding and growing become disjointed. Life becomes harder for plant and animal life. ⅓ of all life will face extinction. 40% of the Amazon rainforest will die, and the hot air, lack of rain and increased evaporation will almost halve all existence in the region. This will drastically reduce its ability to regulate the Earth's climate. As the Amazon dies, carbon dioxide trapped in the soil and within plant and animal life will be released into the air. This will further exacerbate the problem of greenhouse gas emissions.

As the oceans warm, they will be able to store less CO_2, releasing it into the atmosphere, increasing the warming effect [16]. In London, UK, temperatures will regularly reach 40°C in the summer. In Delhi, India, 45°C will become the average temperature. The need to cool people, spaces, and infrastructure will increase energy demand. This will cripple energy supply systems, causing power cuts and blackouts. Already, in the last 30 years, 40% of Arctic Sea ice has melted. The process of melting sea ice becomes self-reinforcing. This is because ice reflects heat, while oceans absorb it. As ice melts, there is more ocean surface. This means more heat is absorbed, raising marine temperatures, making the ice less likely to reform. 125,000 years ago, the Earth was 2°C warmer than it is today, and sea levels were 6 metres higher.

Another concern is that once temperatures have increased by 2°C, stopping the warming by another degree, to 3°C, is almost impossible. This is because the processes of warming start to fuel themselves. As oceans get warmer, they release carbon dioxide, increasing the greenhouse effect; the Earth gets warmer. As the Earth gets warmer, the Amazon dies, temperature regulation is reduced, and carbon dioxide is released; the Earth gets warmer. As plants suffer from heat stress, they stop taking in carbon dioxide, leaving more in the atmosphere; the Earth gets warmer.

3°C Higher

At 3°C hotter, plants stop absorbing carbon dioxide and start to die. This will increase global temperatures by another 1.5°C. If this happens, by 2100, we will experience runaway global warming. Hurricanes will be stronger; Australian, Asian and South American cities will be destroyed. Rivers will run dry, with no snowmelt from the mountains left to feed them [17]. With no rivers, there will be less freshwater, less arable land for farming, and seawater will flow up where rivers once flowed down. The ground will be poisoned at the least, and completely covered by seawater at the worst [20]. For every degree over 30°C, grain yields drop by 10%. Agriculture, in many regions, will move northwards to find suitable climates for growing. Yet, droughts and unpredictable rain will make this extremely challenging to manage [21].

3,000,000 years ago, when the planet was this hot, sea levels were 25 metres higher. If that were the case today, the Netherlands would be destroyed by the North Sea, and trees would grow in the Arctic.

4°C Higher

By the time the Earth reaches temperatures increased by 4°C, there would be no polar ice, and both Bangkok and Shanghai would be completely underwater [22]. Rainforests would have turned into deserts. Temperatures were this high 55,000,000 years ago when alligators lived in the Arctic and mangroves grew in England.

Economies would crumble in this situation, with trade becoming impossible, as no one would be able to produce anything to trade. Under these conditions, the release of methane hydrate would become another driver of global warming [23]. Methane hydrate, also known as methane clathrate, is a crystalline substance. It forms when methane gas molecules are trapped within a lattice of ice. Methane hydrate can only form under specific conditions of low temperatures and high pressures. This is typically found in marine sediments on continental margins and in permafrost regions. Methane hydrate can release methane when disturbed or warmed. At a global temperature increase of 4°C, methane hydrates can release methane, a potent greenhouse gas.

The release of methane from melting hydrates would speed up global warming. This process would create a **positive feedback loop**. Warmer temperatures lead to more methane release, driving further temperature increases. The process will amplify climate change. In the worst-case scenario, large and sudden methane releases could occur. This could trigger rapid and severe global warming events, as **methane is 20 times more potent as a greenhouse gas than CO_2.**

6°C Higher

The effect of methane hydrate becomes catastrophic at 6°C above preindustrial times. Vast areas of the Arctic permafrost store huge quantities of methane hydrate. These regions would rapidly thaw. This could lead to widespread and continuous methane releases from permafrost-bound hydrates [24]. Warmer temperatures would penetrate deeper into the oceans' seabed. This would destabilise methane hydrates in marine sediments, particularly in shallow waters. The thermal expansion of seawater would contribute to a warming ocean column, speeding up the breakdown of these deposits. With such a large temperature increase, disastrous releases of methane become inevitable. These are sudden, explosive, large-scale discharges of methane from destabilised hydrate deposits. Historically, these events have been linked to rapid climate changes. But at this point, the climate is already incompatible with human life. The increased methane levels in oceans would contribute to ocean acidification as methane oxidises into CO_2 in the water. This would further damage marine ecosystems.

Throughout the history of the planet, similar large-scale methane releases are thought to have occurred before. It is believed that they contributed to some of the most dramatic extinction events. This includes the Palaeocene-Eocene Thermal Maximum (PETM), which occurred around 56 million years ago [25]. There was a sudden temperature rise linked to widespread species extinctions. This may have been due to massive methane hydrate releases. A 6°C temperature rise could trigger similarly catastrophic events.

251,000,000 years ago, the Earth was 6°C warmer than pre-industrial times. If temperatures were to increase to this level again, 95% of all species would become extinct [26]. Life as we know it would not be possible in seas and oceans, as creatures and plants would suffocate. Hydrogen sulphide would concentrate in the warmer waters, poisoning anything that could survive. Hydrogen sulphide in the air would poison trees and animals. It would destroy the ozone layer, leaving the planet vulnerable to extreme levels of ultraviolet radiation.

The Future is Already Here

Texas

In February 2021, Texas faced three of its worst winter storms, leading to the deaths of hundreds of people. During February 10-11, 13-17 and 15-20, these storms caused almost complete energy infrastructure failure in the state. Water and food shortages affected many people, as did a complete lack of heat, caused by the collapse of the power grid. More than 4.5 million homes and businesses were left without power, which, directly or indirectly, led to the deaths of at least 246 people [27]. People were found frozen to death in their homes. Others died from carbon monoxide poisoning as they attempted to use the heat from their cars to keep warm in closed garages. The official death toll is claimed to be understated, with some estimates saying that as many as 700 people were killed as a result of the crisis [28].

At the time of the storms, frozen wind turbines and solar panels were blamed for power cuts. State officials, including Republican Governor Greg Abbott, made several statements. He used these renewable energy sources as a scapegoat and a reason to continue burning fossil fuels. Yet, data showed that a failure to winterise power sources caused the power grid to fail. The predominant failure was from the natural gas infrastructure. The power loss from this system was more than five times greater than that of wind turbines.

These storms caused a record low temperature at Dallas/Fort Worth International Airport. A low of $^-19°C$ ($^-2°F$) was recorded on February 16, the coldest in North Texas for 72 years. Texan homes, in general, which rarely experience low temperatures, are poorly insulated. They have heating systems that are inefficient for heating homes in such low temperatures. This caused an extremely high electricity demand, which the power infrastructure in Texas was not prepared for. Previous storms had revealed this weakness in the power grid, but it was not addressed. Governor Greg Abbott later acknowledged that coal, natural gas and nuclear plants had played a role in the failures. The problem was not that solar panels weren't receiving enough sun; the main problem was that power systems need a certain amount of power to function. The power cuts caused by the storm affected gas compressors, cutting off the supply to power plants.

As those affected continued to deal with the aftermath, people blamed renewable energy sources for the power failures. On social media, images of helicopters de-icing a wind turbine were posted, with people continuing to say that wind turbines froze. These social media posts, which went viral, used photos that were taken in 2015 in Sweden, more than 5 years before the Texas storms. The resistance to the truth was strong. Some groups still claimed that renewable energy sources cannot provide enough energy. That they are unreliable and are more vulnerable to weather damage.

Wales

In August 2022, meteorologists stated that the seas surrounding the UK are rising at a far faster rate than a century ago. The head of the Environment Agency warned two months before this that some coastal communities "cannot stay where they are" [29]. The Welsh village of Fairbourne is one of these communities. It is now being called the first place in the UK to be lost to climate change, as rising sea levels and extreme weather become too much to defend it from.

The small community is facing an uncertain future. The local council has said that they are unable to defend the village from the rising sea and that Fairbourne will have to be relocated by the mid-2050s. Where the villagers will be moved to, they do not know. This has left them living under the constant threat of flooding and storms that cause sea surges [30].

Fairbourne sits at the mouth of an estuary, which poses a flood risk. It faces more risks of flash floods from the river running behind it, and increasingly powerful and frequent storms. To the front, there are rising sea levels. Officials have spent millions of pounds strengthening a sea wall and almost 2 miles of tidal defences. But this cannot protect the village from the effects of climate change.

In Great Britain, this is not an isolated problem, with flash floods and other extreme weather events becoming more common. The winter of 2023 was a particularly wet season. Flooding from persistent heavy rain affected farms, impacting food supplies. Public transport and other vital systems were also affected. The UK government now works with meteorologists, anticipating weather patterns and preparing accordingly. This is to ensure that emergency responses will be available across the nation to protect and assist the population. The British Red Cross says that 1.9 million people across the UK, currently live in areas at significant risk of flooding. This number could double by the 2050s, with the cause being attributed to climate change.

Central Europe

In September 2024, Northern Portugal declared a "state of calamity" due to wildfires after a hot, dry spell [31]. At the same time, Storm Boris wreaked havoc across Central and Eastern Europe. The highest flood warnings were declared across 100 areas in the Czech Republic, with most regions in the country affected. Poland experienced some of the worst flooding in 30 years. This caused Polish Prime Minister Donald Tusk to declare a state of natural disaster [32]. Austria and Romania also suffered from severe flooding; the River Danube flooded, reaching a peak depth of 970 cm (382 inches) [33]. The Morava River, which flows south from the Czech Republic to Slovakia, flooded several small towns and villages. Anyone in the affected areas who did not evacuate in time risked being stranded or even washed away. People reported seeing dead bodies floating through the streets, carried away by floodwaters.

Rescue efforts included rescuing people who were trapped by the high water, as well as animals, including dogs and horses. Photos were published of geese, seeking a reprieve from the water, standing on picnic tables. In Poland, the army had to set up field hospitals to help the injured, as getting to medical facilities proved impossible. Storm Boris led to the deaths of 24 people and billions of euros worth of damage.

The unprecedented storm was caused by a low-pressure system that stalled over central Europe. Cold air from the Arctic met with warm, moist air from the Mediterranean. The moisture in the air precipitated and resulted in heavy rain that persisted for several days across central Europe. The rain spanned from Austria to Romania, leading to some of the worst flooding in nearly three decades.

The damage from the floods included streets covered in mud and debris, damaged bridges, as well as burst dams and embankments. People who survived the flooding were affected by power cuts. They had to be provided with food and clean water by trucks across all the affected countries.

This storm, like the one in Texas, is no longer a rarity, as intense storms have been made more likely by the warming climate. Warmer air is able to contain more moisture, which, when mingled with cool air, can then pour down from clouds in immense volumes. EU Crisis Management Commissioner, Janez Lenarčič, has been vocal about the need to address climate change. Storm Boris prompted him to remind people and governments of the reality that we all face.

> *"Make no mistake. This tragedy is not an anomaly. This is fast becoming the norm for our shared future," - Janez Lenarčič*

"Europe is the fastest warming continent globally and is particularly vulnerable to extreme weather events like the one we are discussing today. We could not return to a safer past," Lenarčič told EU lawmakers in Strasbourg, France. "The average cost of disasters in the 1980s was 8 billion euros per year. More recently, in 2021 and in 2022, the damage surpassed 50 billion euros per year, meaning the cost of inaction is far greater than the cost of action... We face a Europe that is simultaneously flooding and burning. These extreme weather events ... are now an almost annual occurrence," he said.

"The global reality of the climate breakdown has moved into the everyday lives of Europeans." [34].

> *Without human-induced climate change… heat like in July 2023 would have been virtually impossible to occur in the U.S./Mexico region and Southern Europe if humans had not warmed the planet by burning fossil fuels. – WorldWeatherAttribution.org* [35]

The Poorest Are Hit the Hardest

The poorest countries, those with the lowest GDP, are often the most harmed by climate change. Droughts, floods, monsoons, and sinking lands are causing devastation to many vulnerable populations. These countries cannot afford mitigation measures [36]. Climate displacements are on the increase, with people being forced to move to safer areas. Many low-income countries rely heavily on agriculture for their economies and livelihoods. The disruption to weather patterns is causing droughts, floods, and crop failures. In 2022, 32,600,000 people were displaced for reasons directly related to climate change, a 41% increase compared to 2008 levels. Not all of these people are moving because of a particular extreme weather event. Many are moving as farming land becomes harder to find. They suffer as water resources dry up, or in coastal regions, sea level rise encroaches on villages and communities.

Poorer nations often lack the money and technology needed to cope with climate change. Infrastructure, emergency, and healthcare services are often under-invested. This makes it harder to respond to climate-related disasters. These countries are unable to invest in climate resilience measures. They cannot afford sustainable agricultural practices, renewable energy, or disaster preparedness programmes.

Climate change also exacerbates existing social inequalities. Vulnerable groups like women, children, and marginalised communities are often the poorest. They face greater risks and have fewer resources to handle climate impacts. These factors create a cycle of vulnerability, making it harder for poorer countries to develop and thrive. Addressing these issues needs global cooperation. Investment in sustainable development and targeted support for vulnerable communities is vital.

An Invitation to Act

The urgency of addressing climate change is undeniable; our future as a species depends on the changes that we make today. Governments and industries are key in enforcing solid policies. They also invest in renewable energy and sustainable practices. But individual actions are equally important. Each of us can contribute to this global effort by making conscious choices in our daily lives.

One of the most impactful changes we can make is adopting a vegan diet. By choosing plant-based foods, we significantly reduce our carbon footprint and pollution levels. Animal agriculture is a leading contributor to greenhouse gas emissions. It also causes deforestation and water pollution. By eliminating or reducing animal products from our diets, we help lessen these harmful effects on the environment.

We must also focus on reducing energy consumption, minimising waste, and cutting down on plastic use. Every day choices, multiplied across communities, can lead to meaningful progress against climate change.

Transitioning to a more sustainable future won't be easy, but it must happen. The decisions that we make now will shape the world that we live in as we grow older and the legacy that we leave for future generations. This journey toward sustainability includes understanding how our food choices impact the planet. By embracing veganism, we can live more mindfully and help create a healthier, more compassionate world.

Together, we have the power to make a difference - **without** compromising our quality of life. By being mindful in our choices, we can help the planet and everything that lives on it thrive for many years to come.

Chapter 2: Why We Now Say Climate Change Instead of Global Warming

A Shift in Terminology

For many populations, summers have become hotter, drier, and longer. Warm weather starts earlier in the year than would be typical and ends later. But, at the same time, winters have become colder and much wetter. Spring and autumn, distinct seasons in the past, are now shortening and blending into the seasons that come before and after. The effect is that plants and animals have become disoriented as the seasons have lost their predictable rhythm. This shift has led sceptics to claim that climate change is a hoax. They say the term "climate change" is simply the rebranding of the previous term, global warming. The planet can't be getting warmer because autumn and spring seasons are colder, and we still get snow in the winter, sometimes a lot of snow.

In the past, the term global warming was widely used to describe the changes occurring in Earth's climate caused by human activities. But scientists and environmentalists began to use the term climate change more frequently. This shift in terminology is not a matter of semantics. It reflects a broader understanding of the complexities of the Earth's climate system. This includes the range of impacts associated with human-induced changes. Although the Earth is warming, the effect is a change in climate around the world and the weather that we experience throughout the year.

Weather: The Forecasts That We Follow

Weather and climate are two different and distinct things. To say that climate change is a lie, based on the weather, is to take a view focused on the minutiae, excluding the overview that gives the weather context. This is like reading one page of a book and believing that you know the whole story. The term weather refers to the short-term atmospheric conditions in a specific place at a specific time. Many elements, including temperature, humidity, and precipitation (rain, snow, etc.), influence weather.

For example, a weather report might say, "Today's weather is sunny with a high of 24°C (75°F) and a light breeze from the west." This describes the conditions you can expect in a particular area during that day.

Also, the weather is a very fluid thing. It is influenced by many factors and can change frequently, leading to variations in the conditions that we experience.

Here are some key factors influencing the weather:

Solar Radiation:

The amount of sunlight a location receives is a primary driver of weather. Solar radiation heats the Earth's surface. This leads to temperature variations that affect weather patterns.

Geography:

Latitude: Locations closer to the equator generally experience warmer weather. The regions near the poles are colder.

Altitude: Higher altitudes tend to be cooler than lower areas. For example, mountainous regions are often cooler than the surrounding lowlands.

Proximity to Water: Areas near oceans or large bodies of water typically have milder weather. Temperature fluctuations are often smaller due to water's ability to store and release heat.

Atmospheric Pressure:

High-pressure systems generally bring clear skies and stable weather. Low-pressure systems can lead to cloudiness, increased wind speeds, precipitation, and storms.

Wind Patterns:

Wind moves air masses around the globe, bringing different weather conditions with them. For example, winds from the ocean might bring moisture and rain, while winds from deserts can bring dry, hot conditions.

Ocean Currents:

Ocean currents can transport warm or cold water across large distances. This can influence coastal weather. For example, the Gulf

Stream brings warm water and milder weather to parts of the North Atlantic.

Humidity:

Humidity, or the amount of moisture in the air, plays a significant role in weather. High humidity can lead to cloud formation and precipitation, while low humidity often results in dry conditions.

Topography:

The physical landscape, such as mountains and valleys, can influence weather patterns. Mountains can block wind and create rain shadows, where one side of the mountain receives heavy rainfall, while the other side remains dry.

Human Activity:

Urbanisation, deforestation and pollution can influence local weather patterns. For instance, cities often create "urban heat islands". These are small areas where temperatures are higher than in the surrounding rural areas. They are caused by heat stored in buildings and structures. Hot air can also be trapped beneath layers of pollution, which is generally worse in cities.

Seasons:

The tilt of the Earth's axis causes seasonal weather changes. For example, during the summer, regions receive more direct sunlight and experience warmer weather. In the winter, there are colder conditions due to less direct sunlight.

Climate Change:

Long-term changes in climate can alter weather patterns. For instance, global warming is leading to more frequent and intense heat waves. This is changing precipitation patterns and increasing the severity of storms.

Global Warming: A Narrow Focus

Global warming is seemingly an inappropriate term given the severe winters that some regions of the world are facing. It is a genuine force that is threatening the future of the planet. Global warming refers to the rise in Earth's average temperature due to greenhouse gas build-up in the atmosphere. This concept gained widespread attention in the late 20th century. In 1988, NASA scientist Dr. James Hansen testified before the U.S. Congress. He presented data showing that global temperatures were indeed rising [14].

Global warming is a straightforward concept. Greenhouse gases like **carbon dioxide (CO_2), methane (CH_4)**, and **nitrous oxide (N_2O)** accumulate in the atmosphere. They trap the heat that would have otherwise escaped into space; this causes the Earth's temperature to rise. This warming has been measured and documented, with records going back decades. They clearly show that the planet's average temperature has increased by approximately 1.2°C (2.2°F) since the late 19th century.

But global warming only accurately describes that the average temperature of the planet is going up. Focusing solely on temperature increases can be misleading. It suggests that the problem is primarily about heat, whereas the reality is far more complex.

Climate Change: A Broader Perspective

Climate change is an accurate name for what we are living through, particularly with the context offered by weather patterns of the past. It is long-term changes in temperature, precipitation patterns, and sea levels. It is also the frequency and intensity of extreme weather events. These changes can manifest in various ways, depending on the region and season. For example, some areas may experience hotter and drier conditions in the summer. Another region might see more intense rainfall or colder temperatures. At certain times of the year, this can manifest as extremely low temperatures or severe snowfall.

"Climate change" also highlights that human activities are altering the entire climate system. This includes changes in wind patterns, ocean currents, and the distribution of ecosystems. The timing of natural events, like plants blooming, harvests and animal migrations, are all affected. By using "Climate change," we can address all the consequences resulting from man-made shifts to the environment.

The Importance of Language in Public Perception

The choice of terminology also has large implications for public perception and policy. "Global warming" often evokes the image of a uniformly hotter planet. This can lead to misunderstandings about the nature of the problem. For example, during unusually cold winters, some people mistakenly argue that global warming is a hoax. It doesn't mean that global warming is the cause of climate change.

As "climate change" encompasses more of the issues that we are facing, the highs are higher, and the lows are lower. We are getting more of everything in ways that are detrimental to our way of life. It helps to communicate that this matter is about much more than rising temperatures. Climate change tells us that the Earth's climate is significantly shifting. These shifts are having diverse and far-reaching impacts on natural systems and human societies.

Furthermore, "climate change" emphasises that the problem is not static. The climate hasn't "changed"; the shift is ongoing and evolving. It's not about past or current warming. It's about the continued and accelerating changes that we will experience in the future if greenhouse gas emissions are not curbed.

Scientific Consensus and the Role of the IPCC

The shift, from "global warming" to "climate change", is also tied to the growing body of scientific research that has emerged over the last few decades. The **Intergovernmental Panel on Climate Change (IPCC) was** established in 1988. It has played a crucial role in advancing our understanding of climate science. The IPCC's assessment reports compile the latest research from around the world. They have increasingly focused on the multifaceted nature of climate change.

These reports have shown that, while global temperatures are rising, this warming leads to a cascade of other effects. Changes in precipitation patterns, sea level rises and shifts in ecosystems are now common. An example is the melting of polar ice. This not only contributes to sea level rise but also disrupts ocean circulation patterns. Changes to ocean circulation can affect weather systems globally. Similarly, warming temperatures can lead to more intense and frequent wildfires. Wildfires, such as those seen in California, release more CO_2, exacerbating the problem.

The scientific consensus, as reflected in the IPCC's findings, is that human activities are driving these changes, and the impacts are likely to become more severe if we do not reduce greenhouse gas emissions. Using the term "climate change" better communicates the broad range of changes and risks associated with this crisis.

Policy Implications of the Terminology Shift

The way we talk about climate issues influences policy decisions. Initially, the focus on "global warming" led to early efforts to limit temperature rise. But, as the understanding of climate change expanded, so too did the scope of policy discussions. Now, it's clear that addressing climate change requires a comprehensive approach. This includes not only reducing greenhouse gas emissions but also adapting to the changes that are already in motion.

Mitigation strategies are crucial to slow climate change. These include transitioning to renewable energy, reforestation, and improving energy efficiency. Adaptation strategies are crucial to tackle the ongoing impacts. This includes building flood defences, developing drought-resistant crops, and redesigning infrastructure.

The shift to "climate change" in policy discussions also highlights the need for global cooperation. Climate change is a global problem. It requires coordinated efforts across nations, industries and communities. Calling this climate change makes it clearer that the challenges are linked, that any solutions must be comprehensive and inclusive. This global stance also highlights how climate change affects different areas of the world. From coasts to deserts to forests and cities. The way that we experience climate change requires a different strategy, depending on the way that the impact is felt.

Media and Communication

The media plays a critical role in shaping public understanding of climate issues. This is where most people learn about the changes taking place, as well as how they are affecting communities and habitats around the world. As scientists have moved from "global warming" to "climate change," media outlets have also shifted. Today, global warming is rarely mentioned. This shift has helped to foster a more nuanced public dialogue about the nature of the problem and the solutions needed. Yet, challenges remain in how climate change is communicated. The complexity of the issue can sometimes lead to confusion or misinformation. For instance, the use of technical jargon or overly simple explanations can either overwhelm or mislead the public. Effective communication about climate change requires balancing scientific accuracy with clarity and accessibility.

Then there is the role that social media plays in informing people of how the planet's climate is changing. Social media plays a significant role in communicating the concept and importance of climate change. It acts as a powerful platform for education, awareness, advocacy and mobilisation. There are several ways it contributes to the climate change conversation. Social media raises awareness. With its global reach, it can explain what climate change is and how it is affecting different populations around the world. Social media platforms can be used to educate the public by sharing research and science in ways that are easy to understand and digest. It also makes activism and the forming of like-minded groups easy. These groups will mobilise. They campaign for governments and companies to make policy changes. They also show us what we can do as individuals to mitigate the threat of climate change.

But with social media comes misinformation and echo chambers. Social media algorithms often create echo chambers. This is a situation where users only see content that aligns with their views. It reinforces their ideas, potentially limiting exposure to diverse perspectives or important information. Misinformation about climate change is a large challenge, particularly on social media platforms. This is due to everyone having the ability to publish their views and opinions to global audiences. It involves the spread of false, misleading, or distorted information, often by those doing the most damage. This can confuse the public and undermine scientific consensus. It also slows the urgent action needed to address climate change [37]. **Climate denialism** rejects that climate change is real, that it is caused by human activity, or that it poses a serious threat. Conspiracy theories claim that climate change is a "hoax". Others state that it is exaggerated by governments or scientists for profit or control. These theories create distrust in legitimate research and institutions. There is also the "information" spread by corporations. Tactics like greenwashing mislead the public into believing that practices are eco-friendly. This is most problematic when no meaningful changes have been made. False solutions may promote ineffective or harmful ideas as being beneficial. Exaggerating the benefits of geoengineering or carbon capture is an example. Especially when the root causes of emissions are not addressed at all.

Combating climate change misinformation requires a multi-pronged approach. This includes scientific education, stronger platform accountability, fact-checking, and engaging communication. Misinformation thrives in environments where people feel confused or mistrustful. Fostering transparency and encouraging informed public discourse is essential. It is also crucial to use education, public awareness campaigns, and responsible journalism. These ensure that people understand the full scope of climate change. They will understand the truth and can discern the lies, motivating them to act. The terminology that we use is part of this effort to engage the public in meaningful and informed discussions about the future of our planet.

Climate Change Rebranded?

Arnold Schwarzenegger, bodybuilder, actor and former governor of California, has campaigned on climate change for over 20 years. Through these years, he has seen little positive change, and he has experienced a lot of pushback. Climate change wasn't considered important. People believed that the effects of climate change were being exaggerated. They refused the idea that human activity has caused climate change or that it is simply a con. He believes that the best way to combat this is to rebrand climate change, to call it pollution [38].

Schwarzenegger believes that traditional environmental messaging is too removed from people's lives. That it fails to connect with people's immediate concerns. By focusing on the issues people care about - health, jobs, the economy, and practical solutions - he aims to make the issue more personal. He also seeks to make the message urgent and politically neutral.

There are several key points to the rebranding project. He wants to frame climate change as a health crisis, publicising the lethal effects of pollution and avoiding the use of the term "climate change". But he also highlights the economic benefits of clean energy and non-political, positive messaging. Importantly, Arnold Schwarzenegger understands the connection between animal agriculture and climate change.

> *"I stopped eating meat because raising cattle (livestock) creates more pollution than all transportation combined." - Arnold Schwarzenegger*

Schwarzenegger has become an advocate for plant-based diets. He starred in and was an executive producer for the film Game Changers, a film showcasing vegan athletes. He also continues to lift weights and says that he feels healthier. He has stated that his cholesterol levels have fallen to a healthy level after being high when he consumed a diet heavy in animal products.

Arnold Schwarzenegger has used his platform to broadcast this message. As a former governor, he has experience with environmental and renewable energy policies. This has made him a credible advocate for this rebranding.

By making it an everyday problem (air quality, health, the economy), Schwarzenegger hopes to engage people. The idea of changing the name of climate change to pollution to raise its public profile and gain support for the need to act is positive. But climate change is about more than pollution. It is the abuse of the Earth's resources, the genuine shift in climate patterns, and the pressure that this puts on ecosystems and the environment. Take the example of rising sea levels; this phenomenon is not explained by "pollution".

A Better Name

Maybe it is time for a new name. One that reflects the seriousness of our situation. Rebranding the term "climate change" could be the next step in getting the public, including the sceptics, to recognise that this is urgent. We are already seeing real and immediate impacts on our environment. Animals and people are dying. Entire ecosystems are being wiped out. Some communities are facing the prospect of becoming climate refugees soon; one extreme storm could wash their lives away. The term "climate change" sounds neutral and gradual. It can, and has, led to complacency, and this has made it easier for people to disengage from the problem. By adopting a more impactful term like "climate crisis," we can convey the severity of the issue and the need for immediate action. This shift in language would emphasise that we are not simply experiencing a natural change in the climate. That we are facing a crisis that demands solutions and accountability.

> *"As scientists, we urge [the] widespread use of the vital signs and hope the graphical indicators will better allow policymakers and the public to understand the size of the crisis, realign priorities, and track progress." - Dr Thomas Newsome*

The term "climate crisis" is already used by some scientists, such as Dr Thomas Newsome of the University of Sydney [39]. It carries emotional weight, which is crucial for spurring public and political action. Reframing the issue helps to make it more personal and pressing. It pushes individuals, companies and governments to recognise the seriousness of the situation. In marketing and branding, using language that provokes an emotional response is a proven strategy. People are more likely to react and make changes when they feel a personal sense of urgency or danger. A term like "crisis" effectively communicates this. This rebranding could catalyse more widespread support for policies that mitigate environmental damage. It could also encourage individuals to adopt more sustainable practices.

In advertising, companies will rebrand or tweak language and packaging. Slogans will be adjusted to reposition a product or service to appeal to new emotions or markets. For example, products marketed as "eco-friendly" or "organic" tap into concerns about health and the environment. A change in wording can change consumer behaviour. Similarly, the rebranding of terms like "global warming" to "climate change" in the past aimed to broaden the tone of the discussion. Now, intensifying the language to "**climate crisis**" could provoke the stronger action needed. It uses the same psychological techniques that advertisers use to create urgency around a product or issue.

Conclusion: The Evolving Understanding of Our Planet

As we move forward, it is a must to continue expanding our understanding of climate change and its impacts. Communication must be prioritised, as we must stay informed of the latest scientific findings. We must commit to supporting policies that address the root causes of the problem. We must make personal choices that contribute to a more sustainable future.

In the chapters that follow, we will explore how specific industries and practices contribute to climate change and environmental degradation. By examining these issues through the lens of veganism, we will see how adopting a plant-based lifestyle can be an immense help in mitigating these impacts and contribute to the preservation of our planet for future generations.

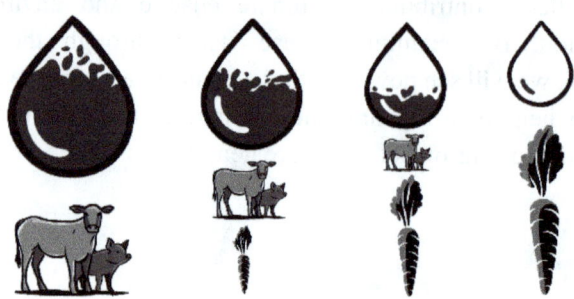

Chapter 3: Reducing Water Pollution Through Veganism

An Introduction to Water Pollution and Agriculture

Without water, there is no life. Water makes up the oceans and seas. It drives the weather and regulates the Earth's temperature. Water sustains ecosystems. Without water, particularly clean water, we have nothing safe to drink and no food to eat. Water is a part of us all and is vital to our existence. Human activities, especially in agriculture, have caused serious water pollution. This is posing a serious threat to both environmental and human health. The farming of animals for meat, dairy and other products is a major contributor to water pollution. This is primarily through the runoff of pesticides, fertilisers and waste products into waterways. Agriculture is also the largest consumer of global freshwater resources. Using about 70% of all freshwater withdrawals [40].

The Impact of Livestock Farming on Water Resources

Livestock farming is one of the largest contributors to water pollution worldwide. This includes farming for meat, dairy, eggs and materials used for fabrics [19]. The impact that this industry has on water resources can be broken down into several key areas:

- Nutrient pollution from manure
- Pathogen contamination
- Animal agricultural waste polluting water streams
- Chemical pollution from pesticides and fertilisers
- High levels of water use

Dead Zones

Animals are living creatures, and they, like us, excrete. When you have a large number of animals, you will have a large amount of excrement. Animal agriculture produces vast quantities of animal excrement, or manure. Manure contains high levels of nitrogen, phosphorus and other nutrients. This manure is often used in fields as fertiliser. It is sometimes used as a dry mix, other times as a slurry, manure mixed with water. This has led to several very serious outbreaks of food poisoning, as manure is not a sterile product. Farmers must follow guidelines set by governments and environmental agencies because manure can exceed the soil's capacity to absorb the nutrients. This means farmers must use it responsibly to prevent excess runoff into nearby water bodies. When waters become too nutrient-rich, a process called **eutrophication** can start.

Eutrophication occurs when an excess of nutrients in water causes **algal blooms**. This is a rapid growth of algae that overpopulates the water. These blooms use up oxygen in the water, leading to "Dead Zones," where aquatic life cannot survive [41]. One of the most well-known examples of a Dead Zone is in the Gulf of Mexico. In this region, nutrient runoff from agriculture, further north in the U.S., has created a large area devoid of marine life. Eutrophication is one of the main causes of the death of coastal ecosystems. Excessive amounts of nutrients in water, caused by the runoff from agriculture, cause plants and algae to grow out of control. This process would, in a healthy body of water, be self-limiting. Algae would grow with the nutrients available in the water. It would then become food for water organisms, creating a natural process of recycling.

Algae is not inherently the problem, but an important part of the food web of healthy water systems. It takes nutrients from the water as it grows and turns them into lipids and other compounds that small organisms need. These small organisms eat the algae, which are then eaten by larger organisms. This process continues, producing nutrients for more algae as creatures in the water die. Their bodies break down, providing food for the algae and other plant life. Eutrophication is not caused by normal algae growth; it is the algae bloom that is the problem. Also, eutrophication can occur naturally as sediment shifts in coastlines and lakes. This changes the character of the water, allowing greater algae growth. Water can become too warm and too clouded for sunlight to pass through. This can create an environment perfect for algae and some plants, but no other forms of life. This is a natural process that takes many years, even centuries. Eutrophication, caused by agriculture, can be a process that occurs in months. It is also made worse in warm weather.

This effect was first identified during the 1960s and 1970s. Scientists found that nutrient enrichment leads to algal blooms from human activities. These included agriculture, industry and sewage disposal

[42]. The estimated cost of damage caused by eutrophication in the U.S. alone is $2.2 billion per year. But the damage caused by eutrophication goes beyond dead zones. Water can turn toxic with foul-smelling phytoplankton. This makes the water cloudy and lowers its quality. The clouded water hinders the growth of green plants, cutting off the source of food chains. As plants and animals die, the water becomes yet more polluted by the decaying remains. The increased photosynthesis of algae uses up the inorganic carbon dissolved in water. This raises the pH levels during the day as temperatures rise, making the water extremely alkaline.

When these dense algal blooms die, the cycle of decomposition begins again. Oxygen dissolved in the water is consumed, leading to **hypoxic** (low oxygen) or **anoxic** (no oxygen) dead zones that cannot support life. Some algal blooms can produce harmful toxins. These toxins can poison pets, wildlife, and even people if they swim in or drink the water. Toxins in the domestic water supply change the water's taste. This makes people aware of the issue and can lead to people feeling the water isn't safe. That the water poses a potential risk to public safety.

The Gulf of Mexico Dead Zone

The Gulf of Mexico is a large body of water, bordered by the United States to the north and east, Mexico to the west and south and Cuba to the southeast [43]. It's part of the Atlantic Ocean and is one of the world's largest and most important gulfs. The area is home to fish, shrimp, crabs, dolphins, sea turtles, and coral reefs. This area is also crucial to migratory birds, such as the American white pelican (*Pelecanus erythrorhynchos*) and the rufous hummingbird (Selasphorus *rufus*). Estuaries, marshes and mangroves thrive in the warm waters. It is a lush and beautiful area.

The Gulf has, in the past, experienced significant pollution caused by oil spills from oil drilling. The most notable being the 2010

Deepwater Horizon oil spill. This caused extensive environmental damage. The *Deepwater Horizon* oil spill (also referred to as the "BP oil spill") occurred off the coast of the United States in the Gulf of Mexico [44]. An explosion on the Deepwater Horizon oil platform released hundreds of millions of gallons of oil into the water. The oil that reached the coast contaminated the land. It coated birds and other wildlife with toxic oil mixtures and was even found combined with sand, creating a mixture that was hard to clean up. In the water, dolphins and other marine animals died in record numbers. Tuna fish developed deformed hearts and other malformed organs [45]. Residents and clean-up workers also faced health issues from the toxic chemicals [46]. Recently, the Gulf of Mexico's marine life has faced a growing threat. This comes from a Dead Zone that keeps expanding.

The Dead Zone in the Gulf of Mexico is a large area of water, the size of Delaware, USA, that is severely hypoxic. This makes it almost impossible for most marine life to survive there. This dead zone is the result of nutrient pollution from agricultural activities. It is also one of the largest dead zones in the world. Nitrogen and phosphorus from fertilisers, animal manure, and sewage runoff flow south. This is especially true for runoff from the Mississippi River Basin [47]. This region covers a huge area, including much of the U.S. Midwest. This is where crops like corn and soya are heavily farmed and fertilised with manure. The runoff from these chemicals concentrates in the Mississippi River as it flows through the states of America. It collects the runoff from fields and empties into the Gulf of Mexico, fuelling the rapid growth of algae. The eutrophication causes algal blooms, which create thick layers of plant material in the water. When the algae die, they sink to the bottom of the Gulf and decompose. This uses up vast amounts of the oxygen in the water, particularly near the seafloor. Unable to survive in the low-oxygen environment, creatures that were previously part of the ecosystem of the Gulf die or leave the area. This includes fish, crabs, shrimp and other organisms.

The size of the dead zone fluctuates each year. It grows and shrinks depending on the amount of nutrient runoff and weather conditions.

On average, it spans about 6,000 to 7,000 square miles (15,500 to 18,000 square kilometres). Though it has reached sizes larger than 8,500 square miles in some years. Reducing, or even eliminating, the Dead Zone will require a long-term reduction in nutrient pollution. This would allow nature to restore the region. Governmental bodies continue to put regulations in place. They set instructions on the handling of animal waste, but not having animal waste in the first place would be ideal. Restoring the Gulf of Mexico could occur as a natural process. Nature can heal itself, but environmental agencies, NGOs, and locals can speed up this process. By working together, they can help reverse the damage that has been done.

Manure In Your Drinking Water

Animal faeces in your drinking water sounds unconscionable. Unfortunately, there are many examples of manure contaminating water supplies. Excess manure from muck spreading, poor storage, or heavy rain can wash pathogens into water supplies. These pathogens include bacteria, viruses and parasites such as worms from animal excrement. This pollution poses a significant risk to human health. It can lead to waterborne diseases such as E. coli and Salmonella infections. This is a particular problem when they contaminate water supplies used in the food industry. A study, authored by Giannis Koukkidis et al., found that bagged salads are a risk for Salmonella, E. coli and Listeria poisoning. Contamination during salad production causes this issue. It can happen at any stage: planting, growing, harvesting, or processing [48]. The bacteria from animals can live on plants or inside plant structures. In soil or water supplies, it can enter plants through the roots, including bacteria from manure. These bacteria cannot be washed away as they protect themselves with a biofilm. Bacteria will form these protective layers on leaves or the inside of a salad bag. This has led to the recall of many salad products, advisories to not eat romaine lettuce, food poisonings and several deaths.

Animals can detect manure contamination [49], but humans cannot. This leaves us all vulnerable to the dangers that result from spraying manure over fields of food grown for human consumption. The risk is well-known, studied and documented, yet the practice continues.

A myriad of pathogens may be found in manure, which in the presence of appropriate hydrological drivers (storm, rainfall or irrigation event) and pathways, can be transferred into water (Chuah et al., 2016; Buckerfield et al., 2019). This has important implications for human and ecosystem health, food safety, environmental security and trade (Roberts et al., 2016). Pathogens originating from animal wastes or manure application... may be introduced into water resources via several routes. Some of these include through direct faecal deposition of faeces on agricultural soils by intruding wild/feral animals, pastured animals or grazing livestock, runoff/leachates from biologically amended soils, feedlots, animal rearing facilities and manure storage systems, amongst others (Gagliardi and Karns, 2000; Goss and Richards, 2008; Sterk et al., 2016; Riley et al., 2018a,b) [50].

Farmer Fined

In July 2024, a UK farmer, Neil Dyke, was fined after taking water sample bottles from an Environment Agency officer investigating a report of pollution at his farm [51]. The pollution was said to be in a river nearby, so he tried to stop the water from being tested.

At the time of the visit, the Environmental Agency Officer saw a tractor spreading slurry in the next field. He also noted that the water in the brook, close to the field, was a dark brown/green colour. The officer noticed foam on the water and a strong smell. These, the water colour, foam and smell, are classic indicators of slurry pollution. The officer took photos, a water sample, and tested for ammonia. The test showed about 10 mg/L, which was the highest reading his equipment could measure. This level also exceeded permitted limits. As the officer continued his testing, a tractor arrived. The officer explained his visit, but the driver of the tractor disputed the officer's authorisation to take samples. He refused to allow the officer to take samples off-site, then took them from the ground and drove away with them in his tractor. The court ordered Mr Dyke to be fined £289 and pay a £116 victim surcharge and costs of £3,699.

Farmer Given Community Service for Manure Pollution

In May 2024, authorities charged UK farmer Derek Dyer in court with contaminating a private water supply and polluting a stream. The contamination resulted from the collapse of a slurry store, built from farmyard manure [52].

In January 2023, Environment Agency officers investigated reports of pollution in a tributary of the River Isle. They went to Crawley Farm in Yarcombe, near Honiton, after several members of the public raised concerns. The pollution traced back to a large structure that had been storing slurry, where one wall had partially collapsed. Investigators found that the farmer had constructed the structure from farmyard manure. The damaged wall resulted in the entire store's contents pouring out. It caused widespread contamination of the soil, water and other natural areas.

The slurry flowed across two fields and down the hillside, into a wooded area. The volume of the slurry and the speed with which it flowed were enough to damage two wire fences. It left a track that was up to 20 m wide in places. There were 12-inch tidelines on fence posts from the manure before it flowed into the wooded area, over 400 m from the store location. The slurry mixture, intended to be fertiliser, caused extensive damage and contamination in the wooded area. E. coli-infected freshwater springs that supplied drinking water to many houses. Total coliforms were above safe drinking water standards, so residents were given bottled water. The wild swimming pond at a nearby glamping site had to close due to pollution from the manure mix.

The farmer, Mr Dyer, received a community order of 60 hours of unpaid work. He was also ordered to pay costs of £15,388.40, with a surcharge of £114 by the judge.

A UN Review

In 2018, the Food and Agriculture Organization of the United Nations in Rome and the International Water Management Institute published a review of water pollution from agriculture. It explored how bigger populations, changing tastes and intensified farming were causing severe water pollution [53].

Globally, in 1961, the average amount of meat consumed per person in a year was roughly 23 kg. By 2012, this had risen to more than 43 kg per person. These changing appetites are being seen throughout the world. But in countries that ate a more plant-based diet, such as

India and China, meat consumption has increased by more than 50% and 150%. This meat-based dietary approach is very resource-intensive. It requires more land, water and fertilisers. For example, a meat-based diet requires up to three times more phosphorus per year per person than a vegetarian diet. A vegan diet requires less still. Another example of changing diets is the demand for cereals and grains. Despite population growth, the demand for cereals in human diets is not likely to rise much in the next thirty years. Predictions are that the total demand for cereals will increase markedly to supply the needs of meat producers. As a result, estimations state that global crop demand will increase by 70% to 110% by 2050.

This resource-intensive industry is also one of the most polluting. Animal agriculture is one of the top three contributors to the most severe environmental problems, including water pollution. The damage caused by the livestock industry is present at every scale. From the small local farmers to the largest factory farms. The 2018 review stated that the livestock sector is presumably the largest source of water pollution. The major sources of pollution are animal waste, uneaten feed, land used for feed crops and chemical pollutants from tanneries.

Towards the end of the review, the policy suggests several responses. The first is that we, globally, change our diets:

> *Different diets have different environmental footprints. An increase in demand for food with large environmental footprints, such as meat from industrial farms, is contributing to unsustainable agricultural intensification and water quality degradation. - The Food and Agriculture Organization of the United Nations*

Untreated Waste

Farming and agriculture in general have always been rural occupations. Individual farmers would raise animals and grow crops on their land. They would sell their produce to the surrounding community. Starting in the 1940s, agriculture has become an industrialised enterprise. The potential to make a lot of money has led to corporate entities entering the agriculture and livestock business. As they have taken over, the number of small-scale and family-owned farms has declined. This industrialisation has resulted in a decrease in the number of farms, but an increase in the size of operations. In the United States, the number of farms fell from approximately 6.8 million in 1935 to 1.8 million in 2024. At the same time, the average farm size grew from under 155 acres to 434 in the same time frame [54]. This process has enabled the number of animals farmed to increase at a rate greater than the acreage of farms. They are also taken to slaughter at a younger age. In the 20 years from 1982 to 2002, the number of animals farmed in industrial production facilities increased by nearly 246%. The total number of livestock raised in the year 2000 was equal to that of the previous 80 years combined.

A concentration of farm animals in a confined area of land is called an **animal feeding operation (AFO)**. A **concentrated animal feeding operation (CAFO)** is where animals are confined, with no grass or vegetation, for 45 days in a year. These farms will contain more than 100 animals at any given time. The animals remain in indoor stalls until going to a processing plant for slaughter. Keeping this many animals produces a huge amount of waste. In the USA confined livestock and poultry animals in the United States produce twice as much waste as the human population. But where the disposal of human waste is regulated, animal waste is not treated in the same way [55].

Regulations state that human waste must be treated in municipal sewer systems. But these rules do not apply to the waste from animal farming. The waste produced by CAFOs, or factory farms, is commonly stored in pits or in open ponds called **lagoons**. These lagoons and pits are not part of a treatment process for the waste; they are a way of containing it. These collection areas are large holes dug in the soil and are often lined with just a layer of clay before being filled with waste from the farm. They often leak their contents into the soil and waterways, leading to nitrates, animal medicines, pathogens and organic wastes polluting surrounding areas. The waste generated by CAFOs will contain animal excrement, bedding, and floor coverings such as straw. Cleaning solutions and sometimes dead bodies can also be found. This mixture will sit in a fetid pool until disposed of. This is sometimes done by treating the waste as slurry. Other times, the disposal means filling the hole with soil. Waste that isn't disposed of remains in the pit until it is full, at which point a new pit will be started. A hole is dug and prepared before being filled with the waste, decay and rot that is left over from animal farming.

Blue Baby Syndrome

Animal waste in water supplies has a direct impact on human health. The water pollution can cause food poisoning, skin conditions and respiratory illnesses. One of the less well-known health impacts is **blue baby syndrome** [56]. Blue baby syndrome can have several causes. One cause is water pollution from agriculture. When a pregnant woman drinks water with high nitrate levels, her baby may produce methaemoglobin. This is an oxidised form of haemoglobin, not the healthy foetal haemoglobin needed. This oxidised haemoglobin cannot bind to oxygen. Soon after birth, the baby will appear blue as they are no longer able to receive oxygen from their mother. This can be fatal in infants if they are not treated, as they will suffocate.

Exhausting Fresh Water Supplies

This issue goes beyond what is in the water, as how much water is available to us is an increasing concern. Agriculture requires a lot of fresh water, often using mains or municipal water. This is the same supply that is used for residential and commercial use. Provided by a water company, it can be used for irrigators and other equipment. Natural water sources, such as rivers or wells, provide untreated water. It can also be harvested or recycled by collecting rainwater or reusing water. Regardless, it is all coming from the finite amount of freshwater that is available for human use.

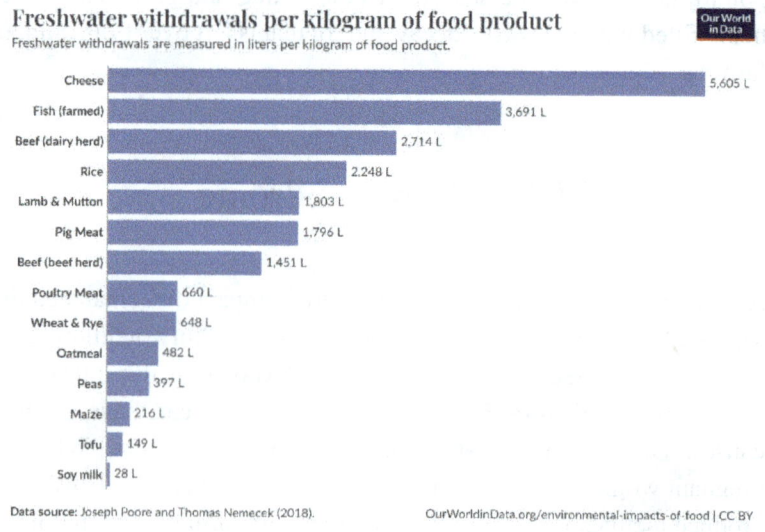

Figure 1

Most of the water used for livestock returns to the environment in the form of manure, slurry and wastewater. This is not water that is safe or immediately usable. This water is hazardous to marine and land animals, including humans [53]. When livestock is concentrated, as in

CAFOs, the waste produced exceeds what nearby ecosystems can handle. This leads to runoff and leaching of contaminants into groundwater and waterways. Figure 1 shows the litres of freshwater needed to produce one kilogram of various foods. Animal products need much more water than plant-based foods to produce. The only exception is rice, as it grows in flooded fields. One of the arguments that vegans often hear is that, if the world were to go vegan, there would not be enough food for everyone. We would need more land to grow enough vegetables, beans, fruits, and grains for everyone. This could deplete our resources. Yet, the truth is that with a vegan diet, we can grow enough food to feed the world, even as the population of the planet grows closer to 8,200,000,000.

Figure 2 shows how much land, by the square metre, is required to produce 100 grams of protein per food source. It shows that, generally, animal-based sources of protein need more land than plant-based sources of protein. The exception would be poultry meat and eggs. This is due to the cramped and overcrowded conditions in which birds are kept. It should be noted that even the farming of fish uses more land than tofu, maize, wheat and rye. To argue that having everyone go vegan would still lead to food shortages is to deny the obvious. Animal agriculture uses land in several ways. Arable land is needed for farms to grow crops for animal feed. Animal feed processing plants, livestock farms, slaughterhouses and meat processing facilities all have a footprint. Figure 3 shows how land use is split. 45% of habitable land is for food production. This equates to 48 million km²; of this, only 16%, 8 million km², is used for growing plant-based foods for humans. 80%, 38 million km², of the habitable land is used for livestock, yet this incredible land use only provides 17% of the world's calories.

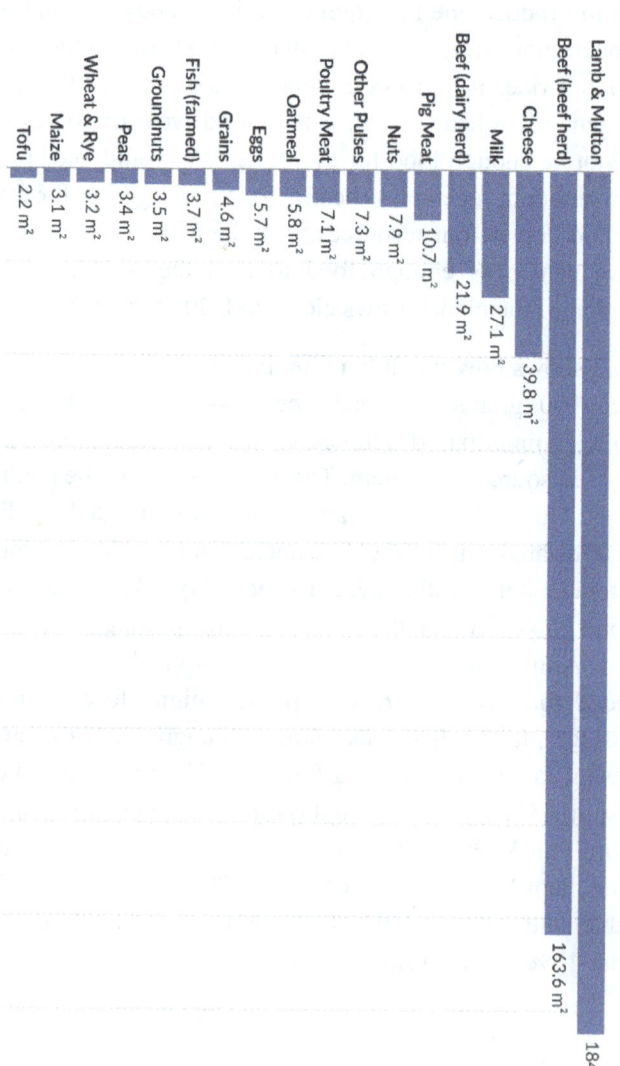

Figure 2

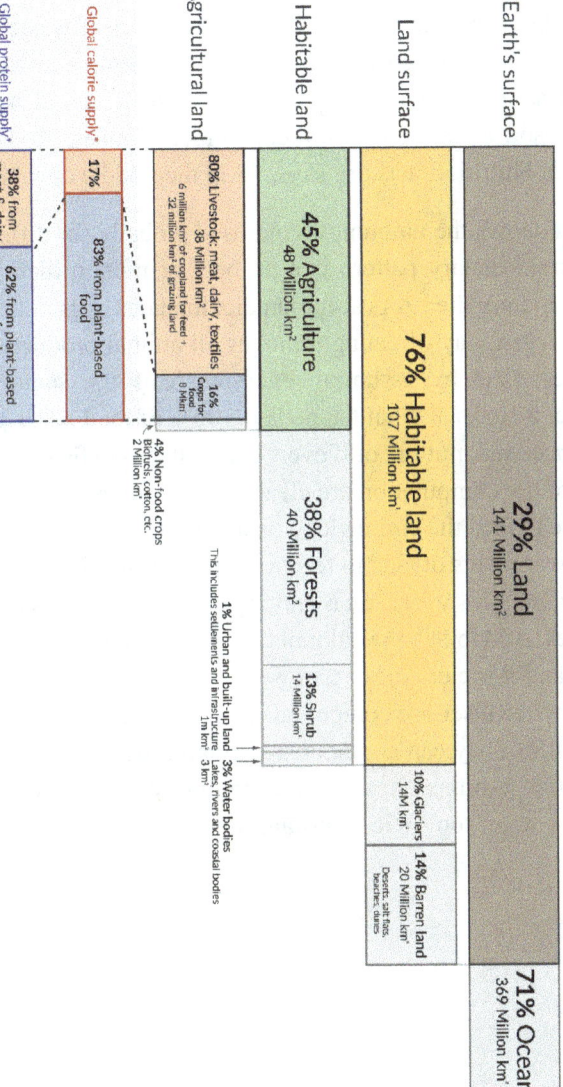

Figure 3

Not all the land used for livestock would be suitable for growing crops; the land may be less fertile and too dry. The land may be more tundra-like, hard and rocky. There is also a lot of waste in feeding crops to animals. In growing a whole living being, only to kill them and use some of their body parts for food. It would be far more efficient and effective to use them to feed humans. As Dr Michael Greger of Nutritionfacts.org says, "skip the middle cow."

Figure 4 shows the amount of land used globally for different diets. The current dietary pattern is a no-beef or mutton diet, a no-beef, mutton or dairy diet. A pescetarian diet includes eggs but no dairy. A wholly vegan way of eating removes all animal products. The data compares diets by assuming everyone eats the same way. For example, it looks at what happens if 100% of the Earth's population avoids beef and mutton, or if everyone switches to fish and eggs. The results of the computation are clear. There is more than enough land available to feed the entire globe on a plant-based diet. Continuing to use large amounts of land to feed and grow farm animals corresponds to water use. This water use is leaving many populations facing water shortages. Add to this the amount of pollution that animal agriculture causes, and it is clear to see that we are heading into disaster. Runoff caused by fertiliser use compounds this. These chemicals are toxic to aquatic life. They can contaminate drinking water sources, posing a serious risk to human health. Like manure contaminating water, this can leave water too toxic to sustain life.

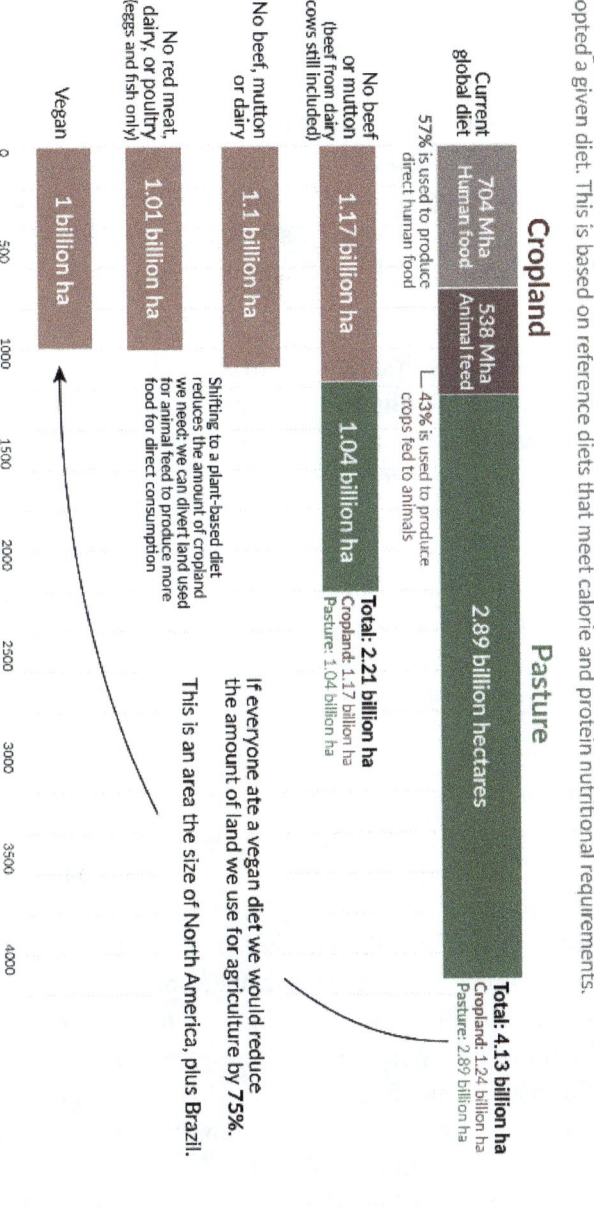

Figure 4

Rules, Regulations and Fines

To stop water pollution from farming, countries have made rules for the use of potential pollutants. This includes clear instructions on how much, when, and where to use them. This applies to both animal and chemical-based fertilisers, and it details the amounts to apply [57]. Inspections make sure that farms follow the rules. Soil and nearby waterways are checked through visual and chemical testing. Non-compliant farmers will face financial penalties and possible convictions.

Buffer zones, planted with shrubs and trees, help protect water bodies by reducing runoff of fertilisers and pesticides [58]. Governments will mandate or encourage the use of nutrient management plans. These plans set out when fertilisers can and cannot be used, such as not using slurry before, during, or immediately after heavy rain. They also optimise fertiliser use to prevent over-application, which can lead to runoff into water sources. Many countries offer financial incentives for farmers adopting environmentally friendly practices. They encourage reduced tillage, precision and organic farming with subsidies or payment of costs [59].

Countries regulate agricultural water pollution through a combination of legislative frameworks and policies. Reducing or stopping manure, slurry, and synthetic fertilisers would improve water quality. These substances are important for crop production. But they contribute significantly to water pollution when management is inadequate. The fact that soil and water must be tested shows that this is not working. This is not a solution to the problem of water pollution. The solution would be to not have the potential for it in the first place. Whether from the spreading of animal waste or the use of non-organic fertilisers. Growing crops on a large scale does need interventions. The ground is prepared, farmers plant seeds or seedlings, and they will be fertilised and, often, watered through irrigation. The plants need to be cared for until the harvest. This is not a natural process, but all this would be scaled down if fewer plants were grown. Less fertiliser would be required if fewer animals needed to be fed to grow muscle to be used as meat. Fewer animals means less manure, less slurry, and less waste. This isn't complicated.

Reducing Water Pollution Through Veganism

All forms of agriculture impact water resources, whether by pollution or use. Plant-based farming has a much smaller impact on water quality and quantity compared to animal-based farming. For example, some plants like almonds and rice are water-intensive. They need large amounts of irrigation, while most, including lentils, oats and peas, need much less. Animal agriculture, the growing of animals for their skins, feathers, milk, eggs or flesh, is, by far, more water-intensive. In animal husbandry, farmers need water to hydrate both the animals and the crops that feed them. They use it to clean their living spaces and to process meat, dairy, and other products produced from them. For example, beef production uses about 15,415 litres of water per kilogram of meat. This is because cows need a lot of water to drink, but also, vast quantities of water are required to produce the feed they consume [60] [61]. Dairy, pork and poultry also have high water requirements. The main problem in agriculture regarding water is eutrophication. Freshwater ecosystems can become contaminated with excess nutrients. This leads to the overgrowth of algae, the death of other organisms in the area and the migration of fish and other marine creatures. The result is the destruction of the ecosystem.

Measured in grams of phosphate equivalents (PO_4eq), it is easy to quantify the amount of eutrophication by food type. In the agricultural industry, animal products are the main cause of eutrophication.

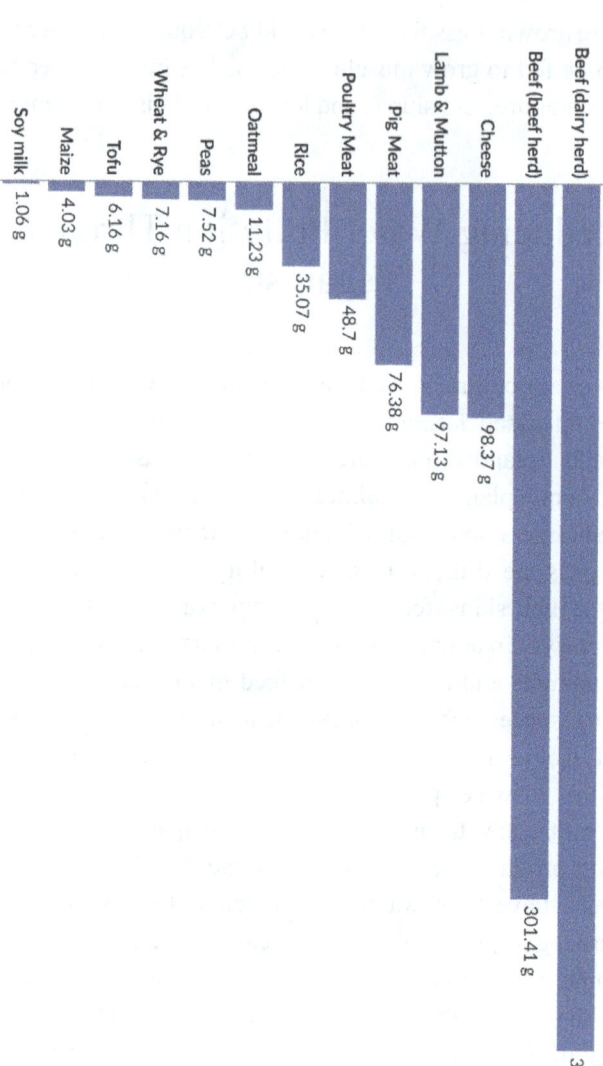

Figure 5

Reducing or eliminating the use of manure and synthetic fertilisers can have huge benefits for water quality. This could seem like a solution: just stop using the pollutant. But, the pollutant, the manure in most cases, is a by-product of another process, animal agriculture. Currently, manure disposal is through spreading it in fields to decompose or fertilise crops. Manure can be burned to generate electricity. When used as cattle bedding, the animals will sleep in their processed manure. It is only in the case of burning the manure to generate electricity that the manure is disposed of. In all other uses, it is moved to a place where it can still be an environmental hazard. Figure 5 shows the eutrophying emissions of different food products by the kilogram. Of the 13 different foods in the chart, the top 6 are animal products. Even with rice being a water-intensive food, it produces relatively little eutrophying emissions. When considering the way that these foods are eaten, it can seem pointless to compare the eutrophying emissions of food such as oatmeal to beef. If you think about swapping beef for tofu in a meal, you could say that picking beef instead of a plant-based option means choosing to pollute water. Figure 1 shows that choosing beef over tofu is choosing to make more water withdrawals. 100 times more. Figure 2 shows that choosing beef over tofu is choosing to use 74 times more land to grow your meal. This isn't an argument for everyone to eat tofu. It is an argument for people to make an informed choice, knowing that they make a big difference.

Comparing Plant-Based and Animal-Based Agriculture

Switching to vegan food can have a profound impact on decreasing water pollution. Reducing the production of animal-based products would reduce land use. It would reduce water use, water pollution and greenhouse gases. Again, these reductions wouldn't only come from having fewer livestock farms. The reductions are also made from not having to grow feed for the animals.

Globally, less than 40% of all cereals grown are used for animal feed. Countries like Spain, Ireland and Denmark feed more than 80% of their cereals to farmed animals [62]. Other crops, like soya beans, are grown in large amounts to feed farmed animals. The rest is used as food for people. All these crops are grown using large amounts of water, pesticides and fertilisers. They are often grown in **monocultures** or **monocrops** and use vast fields of land that could otherwise be left wild. Mono-cropping is where only one type of crop is planted over a large area. This harms soil health due to nutrient depletion and reduced biodiversity caused by the lack of variety in plant species in each area [63]. Growing a single type of plant, like a field of maize, drains the same nutrients from the soil over time. This results in nutrient imbalances and reduced soil fertility over time. Methods such as crop rotation or intercropping protect soil structure. They prevent erosion, nutrient depletion, and enhance water retention.

Plants of different species have differing qualities and requirements. In nature, they will work together to maintain, or even improve, the quality of their environment. Different plants will have varying nutrient needs and root depths. Deep-rooted plants can access nutrients and water from deeper soil layers. Plants with more shallow roots will take water and nutrients that are closer to the surface of the soil. This allows both plants to thrive and not compete for resources. Different root structures in plants can improve soil structure and health. Plants with fibrous roots help to prevent soil erosion. Those with taproots can break up compacted soil and improve aeration. Many plants form symbiotic relationships with soil microorganisms. For example, legumes like beans are hosts for nitrogen-fixing bacteria in their roots. These bacteria convert atmospheric nitrogen into a form that plants can use. This enriches the soil with nitrogen, benefiting nearby plants without the need for manure or artificial fertilisers.

Growing crops in a **polyculture** or **intercrop** system can also control pest populations without the use of chemicals. This works as some plants can deter the pests that will damage another plant when they are grown together. For example, planting basil near tomatoes can help repel pests that affect tomatoes but cause no harm to the basil plant. Growing plants for food in a mixed population of cover crops or ground covers is a more holistic form of horticulture. Low-growing plants, such as clover or vetch, can protect soil from erosion. They will suppress weeds, keep soil cool and maintain moisture levels, contributing to soil health and reducing water use. Growing crops and other plants in intercrops additionally encourages microbial diversity. This is because a diverse plant community supports a variety of soil microbes. These play an essential role in decomposing organic matter, nutrient cycling and disease suppression.

These interactions between plants create a healthier soil ecosystem. They promote fertility, healthy soil structure, as well as disease and pest resilience. This reduces the need for human or chemical intervention. When growing plants in monocrops, there are often issues of pest control. This is because pests and infections can spread through the plant population. Reducing the variety of plants in an ecosystem affects the food chain. This leads to fewer pollinators and other beneficial insects. With only one source of food, insects that prefer other foods will move away. Crop fields can also cause a reduction in habitats for wildlife. Diverse ecosystems are more resilient to changes, such as climate fluctuations, pests and diseases. When planting mono-crops, as is required to produce the enormous volumes of animal feed, all these advantages are lost.

Plant-based diets promote the use of more sustainable and less resource-intensive farming practices. For example, many staple vegan foods, such as legumes, grains, fruits, and vegetables, can be grown using methods such as intercrops. Fields can be designed around plants that have a symbiotic relationship. Farming in this way would prioritise soil health, biodiversity and water conservation.

Water Use

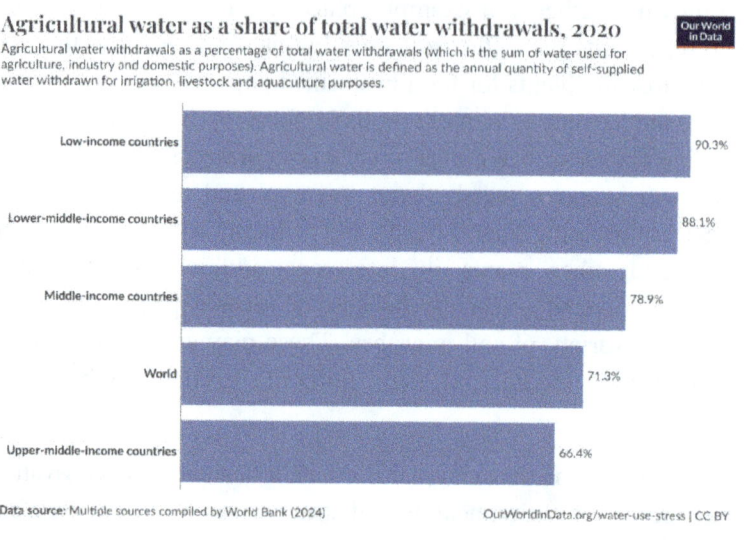

Figure 6

Figure 6 shows the percentage share of water withdrawals, by country, used for agriculture according to their income in 2020. The richest countries are using, on average, 66.4% of their total freshwater for agriculture, while the poorest countries are using 90.3%. Figure 7 shows who in the world is eating the most meat. It's not the poorest countries, yet these are the countries that are most affected by climate change caused by meat production. Unpredictable and extreme weather is making farming challenging across the planet. Poorer regions are particularly impacted, where issues of drought are more pervasive. Wealthier countries, although using less of their water as a percentage, are using far more water in total than poorer countries to produce food. This is due to the production of animal products being so water-intensive. To produce 1kg of beef, 15,415 litres of water are required; 1kg of pork uses 5,988 litres of water, while 1kg of oats needs 482 litres, and corn only

216 litres [60] [61]. The least wealthy countries are relying on grains, root vegetables and other plants to survive. They use more of their water out of the total available to them to grow their nation's food, and their survival is becoming more precarious.

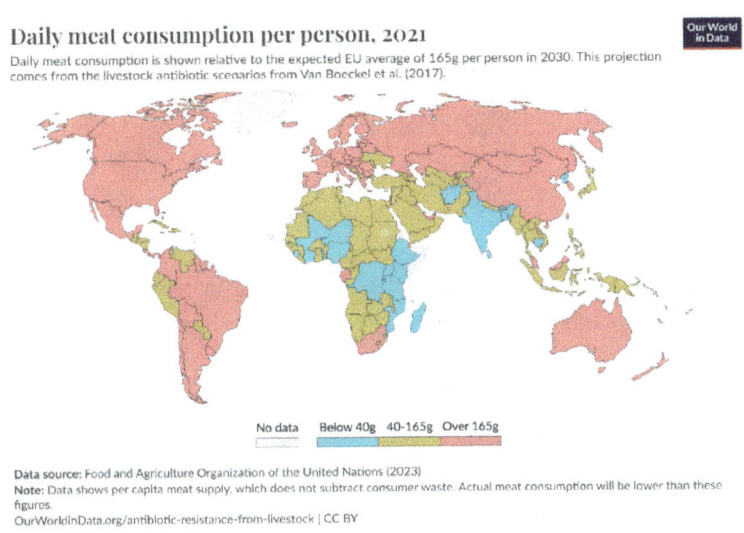

Figure 7

It is important to remember that the planet has a finite amount of water, roughly 326 million trillion gallons. Of that water, less than 3% is freshwater. More than 2% of which is frozen in glaciers, leaving less than 1% for all humanity to use in industry, to drink, wash and grow food with. Analysis by Our World in Data has found that, globally, 39.87% of all cereals grown in 2021 were used for animal feed. Some countries, such as Spain, use far more, with 80.97% of cereals being used to feed farmed animals [62]. Spain is also one of the countries eating the most meat.

The Impact of Soya and Corn Production

It is important to note that not all plant-based agriculture is without environmental impact. Soya and corn production, for instance, have been associated with significant environmental challenges. These challenges are compounded by growing crops in huge quantities to support animal agriculture.

Soya Production

Soya is one of the most widely grown crops in the world. It is a versatile crop and is used in a wide range of food products. But most of the world's soya, more than 77% [64] is used to feed animals grown for meat; just 7% is used to feed people [65]. The expansion of soya cultivation for the animal agriculture industry has led to environmental damage. Regions like the Amazon rainforest have suffered deforestation and habitat loss. The increased use of fertilisers and pesticides has also exacerbated the problem of water pollution.

The environmental impact of soya is vastly different when it is grown for human consumption. When soya is consumed by humans, rather than being fed to animals, the environmental footprint is much smaller. This is because less land, water and other resources are required to produce food for humans rather than to feed animals that will be used for meat. It could be argued that a large amount of the soya used for animal feed is a by-product of producing soya oil for cooking. The bean is pressed, releasing the oil, and what is left, soya cake, is then used to feed animals. But the soya cake that results from soya oil production can be, and currently is in very small amounts, used to make meat substitutes. Products such as meat-free burgers and "meatballs" are made from this soya cake. If animal agriculture were to end, we could reduce soya production by up to 77%. Feeding soya to people, as tofu, meat substitutes, or other soya products, is more efficient and better for the environment. We would also need less soya to get the nutrients that we need if we were to eat the soya ourselves, rather than having it processed through an animal first.

Soya is a nutritious food, being a source of protein, iron, calcium, zinc and copper while being low in fat [66]. The nutrients that people get from eating meat come from the foods that the animals are eating. We can often get these nutrients from the same foods, benefiting our bodies more than by eating them in the form of meat, where many nutrients are lost.

Corn Production

Like soya, corn is another crop grown primarily for livestock feed. Globally, 56% of corn is used for animal feed, while only 12% is fed to humans, with the remaining being used for ethanol, biofuel production and other uses [67]. In the USA, the discrepancy is greater; 1% of corn is grown for human use, with the remaining 99% being used for animal feed, ethanol and other uses [68]. The intensive farming practices used in corn production can lead to significant water pollution. The heavy use of chemical fertilisers and irrigation can mean nutrient runoff and pesticide contamination are inevitable. All of this is reduced if less corn is grown to feed animals.

Corn is also sensitive to heat stress and can suffer quite badly from water stress. During high temperatures, corn plants will close the stomata on their leaves. Stomata are the pores through which gases enter and leave the plant, and water vapour is lost to the air

[69]. This happens to prevent water loss through excessive transpiration. But it also prevents gases from entering or exiting the leaves. This reduces photosynthesis in the plant as carbon dioxide cannot be taken in from the air.

It is an interesting dichotomy that climate change is impacting corn yields, as it feeds one of the main causes of climate change. Corn is not a sustainable crop; it requires large amounts of nitrogen and other nutrients, which it strips from the soil. Corn also needs a lot of water to grow. As temperatures rise, farmers will have to use irrigation to supplement rainfall. The requirements for growing corn result in fields being depleted of nutrients. They can also become dried out, making future crops that much harder to cultivate.

The corn that is fed to animals is not the same variety of corn used to feed humans [68]. The corn we buy in shops, sweetcorn, is softer and sweeter. The corn fed to farm animals is field corn. It is a starchier variety that has a longer growing period, remaining in fields until dried out. Sweetcorn, the variety that we can buy in stores, is harvested when the stalks are still green. Corn, like soya, is a crop that has great nutritional benefits for humans. It is a source of vitamin C, B vitamins and vitamin A, as well as minerals such as magnesium, zinc and calcium [70]. When corn is used in human diets, especially in whole forms rather than processed products, it generally has a lower environmental impact. This is due to less intensive growing requirements. Additionally, corn can be grown using sustainable farming practices. These methods limit greenhouse gas emissions and reduce their impact on water resources. Furthermore, field corn is also used for food that humans eat. When processed, it can be used in products including corn chips and tortillas; it is not a food that is unusable for humans.

If the growing of soya and corn to feed animals grown for meat were to end, the fields used to grow these products could be used to grow other foods. Millet, tomatoes, pumpkins, cucumbers, peppers, sunflowers and beans are examples of plants that grow well together. They require the same conditions and soil type as corn and soya beans [71]. Not only this, but growing these crops in combination with companion planting would have added benefits. Nitrogen-fixing plants, like beans, improve soil fertility for later crops and mixing these crops can help maximise yields and manage pests without chemicals [72]. Growing tomatoes with soya beans benefits tomatoes as soya varieties deter pests that damage tomato plants. Growing corn with millet has many benefits. The height differences between the two plants allow better use of space. Millet can also improve soil quality and suppress weeds, benefiting both crops.

Veganism as a Pathway to Cleaner Water

A significant reduction in animal agriculture would have profound effects on agricultural practices beyond the cultivation of corn and soya beans. We could reduce, or even end, the use of manure in favour of plant-based fertilisers. There would be an obvious reduction in demand for soya and corn crops. Farmers could diversify into crops like millet, vegetables and legumes for human consumption. Along with this come the reductions. Water, fertiliser and pesticide use would drop. This would lead to an improvement in water quality in the water table and other natural water reservoirs. Farmers would pivot to growing a wider variety of crops that are more sustainable and better suited to local conditions. Food security would be improved, and dependency on monocrops would be reduced. As mentioned above, there would be the opportunity for companion planting. But also, more diverse crop rotations and reduced tillage would improve soil structure. Organic matter in soil would increase and enhance natural nutrient cycling. A shift toward plant-based agriculture could promote greater biodiversity. Enabling the cultivation of a wider variety of crops in one space. This, in turn, would support more diverse ecosystems and wildlife.

Planting in this way would also support pollinators, many of which are endangered. By creating a more varied and attractive environment for them, their numbers could recover. Growing a range of plants with different flowering times provides a continuous source of nectar and pollen through the growing seasons. This would attract and support pollinators like bees, butterflies and hoverflies. A range of plants can also create habitats for beneficial insects. For example, planting flowers alongside vegetables can offer shelter and a breeding ground for pollinators. It also enhances biodiversity and helps to increase crop yields. Growing several types of crops together can create "pollinator corridors". These allow pollinators to easily travel between plants, ensuring they find food sources across the entire field.

All these factors can come together to create a better environment for pollinators while improving plant health and productivity. The benefits that come from farming in this way impact more than the environment and local ecosystems. They also impact the farmer, as costs can be reduced while yields increase. The need for less water means less irrigation, a system that needs ongoing outlay for water, running and repair. Fertilisers and pesticides cost money, and getting people to apply them to fields costs even more. With these factors gone, farmers will be better off. The water will be cleaner, the soil safer and the air fresher. Pesticides are toxic, and manure stinks.

Animal agriculture often involves large-scale operations, CAFOs or factory farms that generate significant amounts of manure. This manure contains high levels of nitrogen and phosphorus and a host of pathogens. When washed away by rain or irrigation, they run off into nearby water bodies, leading to contamination and eutrophication. Another contaminant that comes from animal manure is antibiotics given to farm animals. Livestock excrete antibiotics and medications that they are given, much of which is unmetabolised. Leaching and runoff of manure and waste from animal farming can occur from manure spreading. It can also come from manure stores, such as in lagoons or pits, on factory farms. These can release chemicals into water systems, including surface water and groundwater. Antibiotics that end up in the soil from manure spreading can enter plants and crops, getting into food supplies for humans. Meat-processing plants and dairy farms often discharge wastewater containing antibiotics and other pharmaceuticals into municipal water systems. Even after treatment at wastewater plants, many of these compounds persist. This is because conventional treatment methods are not designed to eliminate them.

One of the most concerning effects of this is the spread of antibiotic-resistant bacteria. When antibiotics are present in water, they create an environment where bacteria can evolve resistance to these drugs. These resistant bacteria can infect humans, making common infections harder to treat. This is a growing global health crisis. Resistant infections can cause longer hospital stays, higher medical bills, and more deaths. Humans can get antibiotic residues and other medications from drinking or swimming in contaminated water, or eating fish and aquatic organisms that have absorbed these drugs. The levels of these compounds in water are typically low. But it is currently unknown what effects long-term low-dose exposure could have on human health. They could include hormone disruption, negative effects on unborn foetuses, or allergic reactions.

Some 73% of antimicrobials sold globally in 2017 were for use in animals used for food production. – House of Commons Library

The majority of antibiotics administered in the world are given to farm animals. Antibiotics and other medications in agriculture are used to prevent or treat infections. Farms are crowded environments, meaning that infections can spread quickly. To prevent this, antibiotics are given routinely in the mass dosing of animals or at constant low doses. They can also be used to promote growth in animals. Fewer farmed animals would mean less use of these drugs. Without large-scale animal farming, the amount of antibiotics entering water systems would decrease. Plant-based systems don't need antibiotics or hormones in the same quantities as animals. In crop growing, antibiotics can be used to treat bacterial infections in plants. But the practice is far removed from the systems used to give antibiotics to animals. These include mixing medications with water supplies for chickens. Giving birds a constant, although inconsistent, dose of drugs. A shift to plant-based diets would stop this process. Without animal agriculture's pollutants, the ecosystems in rivers, lakes and oceans are healthier. Plant-based farming, to provide food for humans, has a smaller ecological footprint. It supports biodiversity, it reduces the strain on aquatic life, it means healthier water systems.

The Role of Policy and Consumer Choices

Governments and policymakers have a critical role in supporting more sustainable agricultural practices. This includes providing incentives for organic and sustainable farming. It means regulating chemical use and not subsidising meat farming. It also means investing in the development of water-efficient crops and farming techniques. At the same time, individual consumer choices also drive change. Demand is created by making ethical choices. By only investing in (buying) products that align with this more responsible way of living, demand will be created for products that are better for the planet. By choosing plant-based foods and supporting companies that prioritise sustainability, consumers can reduce the environmental impact of agriculture, promoting cleaner, healthier water systems.

in a single month of being vegan you'd also, on average, save 273 kg of CO_2 emissions, 84 square metres of forest, and 125,000 litres of water. – TheHumanLeague.org [73]

When considering the problems of water pollution, water use and conservation, veganism isn't a panacea. By having the world go vegan, water pollution would not be eliminated. Water pollution is created by many aspects of our lives, from the toilet cleaners we use to the detergent we wash our clothes with. Shampoos and other personal care products are all washed into water systems. There is the impact of industry, manufacturing and the processing of goods, particularly fabrics. The building industry is also a great water polluter, as are petrochemical companies. Some of these problems we can mitigate by changing our habits. We can choose environmentally friendly cleaning products. We can use public transport, ride a bike or walk instead of driving a car, electric or otherwise. We cannot directly impact more industrial processes, but we can, and do, decide what is on our plates and what we eat. The impact that a vegan diet would have on the planet and climate change should not be underestimated. Each vegan person prevents the death of, on average, 1 animal a day, along with all the damage that is done in farming that animal. At the time of writing this book, I have been vegan for almost 9 years. The Vegan Calculator estimates that I have saved 13,407,758 litres of water and the lives of 3,220 animals, and that's just me [74]. I wasn't always vegan. I grew up eating meat, so I understand the transition. Changing how you think and feel about what you eat and where your food comes from. I can see where conditioning and habit keep us stuck, and that is why so much of what we eat is habit. We always have, so we always will; the problem is that we are far too intelligent for that. It is a simple truth that animal agriculture is a clear and present danger to our health, our environment and this planet.

Conclusion

It should be noted that not every person on the planet would be able to eat a vegan diet. Whether the reasons were due to health, geographical restrictions on foods that can be grown, living in a food desert, or any other reason. These populations would still need animal-based foods for sustenance. Eating in this way does not defy the principle of veganism. Despite what many dictionaries and websites define veganism as, it is not only about cutting out animal products. The Vegan Society's formal definition is:

> *"Veganism is a philosophy and way of living which seeks to exclude - as far as is possible and practicable - all forms of exploitation of, and cruelty to, animals for food, clothing or any other purpose; and by extension, promotes the development and use of animal-free alternatives for the benefit of animals, humans and the environment. In dietary terms it denotes the practice of dispensing with all products derived wholly or partly from animals." (75)*

For this reason, eating fish, for example, in an area where this is the only source of protein that would sustain life, would be considered fair. It is more important that animals do not suffer in such circumstances and that they are treated with respect. This idea is not popular with many vegans, but a degree of pragmatism is required when addressing a problem as colossal as climate change.

Vegan diets can reduce many of the environmental problems associated with agriculture. Particularly those related to water pollution, land degradation and greenhouse gas emissions. Shifting to a vegan lifestyle would greatly reduce the mass production of animals for food and other goods. This change would have a significant effect on water resources and pollution. Many countries have already stated that animal agriculture is the main cause, by far, of water pollution. Affecting ecosystems and recreational water uses, such as swimming and water for drinking. By removing this, we can restore water to the clean and beautiful element of our world that it used to be. This would lead to healthier water systems and improved soil conservation. Greenhouse gas emissions would be reduced, and there would be greater sustainability in agriculture. Through more efficient resource use, veganism could help protect biodiversity. It would create a more sustainable food system that benefits both people and the planet.

In the next chapters, we will continue to explore the environmental impacts of different types of animal farming. From the pollution caused by poultry farming to the destruction of marine habitats by the fishing industry. We will see that the choices we make at the dinner table have far-reaching consequences for the planet.

Chapter 4: Pollution, Poultry Farming and the Planet

Poultry farming is a fast-growing part of animal agriculture. The rise in demand for chicken, turkey, and eggs drives this growth, as they are cheap protein sources. Many view poultry farming as less harmful to the environment than beef or pork production. But it still has significant environmental impacts. These include pollution, resource use, and contributions to climate change. Large-scale poultry farming also damages ecosystems and human health. This issue is well-documented and has led to legal actions from communities trying to stop the harm. The damage arises from feed production, transportation, and waste management, and these problems will likely grow as the demand for poultry products increases.

The Scale of Poultry Farming

Sales of poultry products have expanded rapidly over the past few decades. with billions of chickens, turkeys and other birds, raised for meat and eggs each year. Farmers raise billions of chickens, turkeys, and other birds each year for meat and eggs. In 2022, the global poultry industry produced over 123 million metric tonnes of chicken meat and 1.8 trillion eggs. That year, farmers slaughtered 76.25 billion chickens [76]. This vast production requires many resources, including feed, water, and energy. It also creates large amounts of waste and pollution.

Intensive poultry farming, or "factory farming," raises many birds in cycles. The process confines birds, concentrating them to boost efficiency. There are different types of poultry farming. Small-scale farms have fewer than 100 birds. Free-range farms allow birds more outdoor time. Industrial farms can house hundreds of thousands of birds in tight quarters. Organic farming uses feed free from synthetic fertilisers, pesticides, and GMOs. It also bans artificial growth agents like antibiotics. Still, up to 98% of poultry meat and 92% of eggs come from intensive farms [77].

In poultry production, birds have short lives compared to their natural lifespan. Chickens can live 8 to 15 years, but those in the meat and egg industries usually live only 1 to 3 years. This shortened life is due to heart, lung, and joint problems from rapid growth. Broiler chickens, bred for meat, live 30 to 40 days before slaughter. Egg-laying hens live 72 to 80 weeks until they are spent. Ducks, slaughtered at 42 to 56 days old, can live up to 15 years. Those raised for eggs will go to slaughter at around 80 weeks. Geese also have a natural lifespan of 15 years. In farming, they live for 23 weeks for long fattening and 16 weeks for medium fattening. Turkeys, slaughtered at 80 to 140 days, would live more than 10 years in the wild [78] [79].

Broiler chickens are now bred to grow faster and reach slaughter weight in under 40 days. They remain in sheds with 0.06 to 0.1 square meters of space per bird, which means 16 to 20 chickens per square meter. [80]. Egg-laying hens are often kept in battery cages, with less than a sheet of A4 paper, or 0.04 to 0.05 square meters of space, per bird. In barn systems, hens have more space at 0.111 to 0.125 square meters each. Still, this is more crowded compared to free-range systems that provide outdoor access [81] [82]. Welfare standards vary worldwide. In the EU, each bird must have 4 square meters of outdoor space. Indoors, birds have limited space, with 9 to 13 birds per square meter. [83].

While mass production has made poultry products cheaper, it causes many environmental problems. Smaller intensive farms can house 10,000 to 50,000 birds. Larger farms hold 100,000 to 500,000 birds, and larger operations can house over 1 million chickens spread across several buildings. [84]. Farmers replace birds grown for meat every 40 days; the volume of production is enormous. In 2022, the industry slaughtered over 75 billion birds for meat. This means that at least 28.31 billion birds were alive on farms at any one time.

Poultry Waste in Your River

The main environmental issue in poultry farming is waste management. Poultry farms generate a lot of manure. This manure, when not managed according to regulations, can harm the environment.

Feed production and transportation are major contributors to global warming (accounting for 70% of the issue), while manure management is responsible for 40 to 60% of eutrophication and acidification - Leinonen and Kyriazakis [85]

Poultry manure is high in nutrients, especially nitrogen and phosphorus. These are vital for plant growth. Overuse of manure as a fertiliser or storing it in ways not stated in guidelines allows nutrients to seep into nearby water bodies. This seepage leads to nutrient pollution, eutrophication, dead zones and the death of aquatic life. These effects are common around intensive poultry farms.

Poultry manure may also contain harmful pathogens like Salmonella and Campylobacter. These can contaminate water supplies and pose serious public health risks. Pathogens will also pollute drinking water, spreading from poultry farms through runoff into streams, rivers, and groundwater.

> *"We know that there's a very tight link between the tonnage, not milligrams as in human, but tonnage of antibiotics used in agriculture and the emergence of resistance."* - Professor Lindsey Grayson, Infectious Diseases, Austin Health [86]

Antibiotics are a common part of poultry production for successful farmers. Intensive farming requires widespread antibiotic use to prevent disease and promote growth [87] [88]. With birds kept in close quarters, diseases spread at an alarming rate. Intensively farmed birds have no outdoor access, and sheds lack proper ventilation. There is no natural way to prevent sickness in such conditions, leading to antibiotic use. They will be given during initial vaccinations and later often mixed into the drinking water. In many countries, antibiotics also promote growth. There are two theories on how this works. One suggests that antibiotics save energy that birds would use to fight infections. The other claims low doses stimulate cell mitochondria to produce more energy [88]. This use of medications has led to antibiotic-resistant bacteria. These bacteria can spread through poultry meat or manure, contributing to the global public health crisis of antibiotic resistance. When they enter water systems, they can infect humans and animals, making infections harder to treat. People can become very ill with few treatment options [89].

Water pollution is another issue. Poultry farming requires large amounts of water for feed production and control of waste and contamination. It is less water-intensive than beef production, but still pressures freshwater resources. This is especially true in water-scarce areas.

Each chicken needs a constant water supply, especially in intensive farming. The amount varies by age, size, climate, and management practices. A typical broiler chicken drinks about 200-400 ml daily, while layer hens need around 250-300 ml [90] [91]. A farm with 100,000 chickens could need 20,000 to 40,000 litres of water daily for drinking.

Producing feed is the most water-intensive aspect of poultry farming. Most of the feed, by proportion, is grains like corn and soya beans. These crops use large amounts of water, the greatest amount consumed by irrigation. Feed production accounts for 70-80% of the water footprint of poultry farming. For example, producing 1 kilogram of chicken meat uses about 4,325 litres of water. The majority is consumed in feed production [60] [61]. Although poultry has a smaller water footprint than other meats, its water demands are still high.

Water is also needed for cleaning barns, equipment, and managing waste. Cleaning barns and equipment between flocks is intensive and must be thorough. It's crucial to sanitise living areas and all materials used in farming to maintain biosecurity. This includes cleaning transport crates and reusable PPE to reduce disease spread. These processes can use several thousand litres per cycle on a large farm [77]. Once birds go to the slaughter shed, cleaning begins with the clearing of dust and waste. The entire shed is then cleaned and disinfected with chemical solutions and water. This includes cleaning the floor, walls, and ceiling. Equipment is also cleaned and disinfected. All areas are washed with freshwater, with wastewater directed into the municipal drainage system without treatment [92].

The slaughter and processing are a multi-step operation, using a lot of water for sanitation and food safety.

Birds, once captured, travel in crates to the slaughterhouse. Each crate goes through a process of cleaning in automated cleaners that use 200-800 litres of water per hour [93]. For birds, the slaughter process starts with stunning. This usually involves an electric water bath, though gas methods like argon or CO_2 are also used. Hung upside-down by their feet on a track, birds' heads are submerged in an electrified water bath. After stunning, birds have their throats slit and bleed out as they move to the next step: scalding. Scalding requires a lot of water. Submersion in hot water baths, at 50-60°C, loosens feathers for removal. These hot tanks hold between 250 and 350 litres of water [94]. Mechanical pluckers remove feathers before evisceration, the removal of heads, feet, and internal organs. Most steps after hanging the birds are mechanised.

Next, the bodies are cleaned and chilled. Blood, bacteria, and feather removal maintain meat safety. The chilling step is the most water-intensive. Submerging birds in cold water tanks or using air chillers with water misting chills the bodies. Water is also used to clean machinery and working areas, followed by a final rinse of the meat before packaging. Standards dictate water use for processing each chicken. In the UK, the Best Available Technique (BAT) is 8-15 litres per animal. Large operations, processing up to 140 birds per minute, can use 2,100 litres of water every minute.

Air Pollution and Poultry Farming

Poultry farming causes high levels of air pollution. This pollution will be concentrated in the surrounding environment. It damages local ecosystems, but also has a devastating impact on human health.

In many countries, farmers of poultry and other animals do not have to report the air pollutants that they release. This is because air pollution regulations in animal agriculture are often poor. Key air pollutants of poultry farming are **hydrogen sulphide**, **ammonia**, **particulates,** and **endotoxins**.

Hydrogen sulphide is a colourless gas. It is poisonous, corrosive and flammable. It has a characteristic smell of rotten eggs that can spread for miles beyond farming facilities. Hydrogen sulphide damages organs if inhaled or if its salts are ingested in high enough amounts. Symptoms range from breathing difficulties to convulsions and even death. In poultry farming, it is the breakdown of organic matter in manure that produces the gas. When oxygen is limited, like in lagoons or pits, gas production increases. The gas will escape into the air, where wind will carry it away from the farm and over towns and communities [77].

Hydrogen sulphide can cause general symptoms of chemical exposure. Respiratory issues, eye irritation, nausea, and headaches are common. In higher concentrations or repeated exposure, it can lead to severe health problems, such as neurological damage or death. People living close to poultry farms have reported memory loss and trouble processing their thoughts. Health problems that would resolve when they moved away from farms. Long-term exposure, even at low levels, can lead to chronic respiratory illnesses, such as asthma. It is within farms that hydrogen sulphide exposure is greatest.

In 2009, Tyson Foods Inc., the world's second-largest processor of chicken, beef and pork, was ordered to pay $500,000 by a U.S. District Court in Arkansas. The charge was "wilfully violating worker safety regulations, leading to a worker's death". The incident took place in its River Valley Animal Foods (RVAF) plant in Texarkana, Arkansas. In this plant, hydrolysers converted poultry feathers into feather meal with high-pressure steam. The process released hydrogen sulphide gas from the decomposing feathers. It was this gas that killed a maintenance employee working near the hydrolysers. Two other employees, working close by at the time, also required medical treatment [95]. Hydrogen sulphide, as toxic as it is, is not a direct contributor to climate change; it is not a greenhouse gas.

Bird farming produces CO_2 from powering heating, lighting, ventilation, and water systems with fossil fuels. Feed production and transportation also rely on the power from fossil fuels. But there are other greenhouse gases resulting from poultry farming that are far more potent than CO_2.

Methane is a greenhouse gas that can trap **120 times** more heat than CO_2. When manure decomposes in anaerobic conditions, methane can result [96]. Poultry farming emits less methane than ruminant animals like cows. The problem is that large-scale poultry farms generate vast amounts of manure. This results in considerable methane emissions. Another greenhouse gas released from decomposing manure is **nitrous oxide**. Nitrous oxide levels peak under wet conditions and where manure is not stored according to guidelines. It is also a powerful greenhouse gas with **300 times** the warming potential of CO_2 [97].

Ammonia is another colourless gas resulting from the farming of birds, with a distinctive and pungent smell. In vapour form, it has a sharp, irritating, acrid odour. If you smell ammonia, you are at risk of dangerous exposure. Exposure to very high concentrations of gaseous ammonia can result in lung damage and death, as ammonia forms a strong acid in the body. In liquid form, ammonia can cause caustic burns on contact. On farms, exposure to ammonia is an occupational hazard, and workers must protect themselves with the use of preventative protection equipment (PPE) [95].

The main sources of ammonia in agriculture are manure, synthetic fertilisers and the cleaning of bird waste. The decomposition of uric acid in bird manure also produces ammonia. As more manure accumulates, more ammonia is produced. Ammonia is not a greenhouse gas; it has no warming effect on the planet. It is the fine **particulate matter** (PM) that warms the planet. It forms when ammonia combines with air pollutants, like sulphur dioxide and nitrogen oxides. It causes a small but notable amount of warming with a blanketing effect, absorbing sunlight and heating the atmosphere.

Ammonia also dissolves in moisture in the atmosphere. It settles out of the air, descending back onto land or into water, in rain or dry deposits with a low, acidic pH. Acid rain often falls close to where it formed. This means it will fall on communities surrounding poultry farms. At ground level, this causes local soil and water acidification. This kills plants, harms ecosystems and reduces biodiversity, further destabilising natural carbon sinks and making ecosystems less resilient to climate change.

The waste from poultry farming also contributes to air pollution through the production of **Poultry Litter Ash**. Poultry litter is a mixture of poultry manure, feathers and bedding material, like wood shavings or straw. Once cleaned out of sheds, the litter is incinerated at very high temperatures. This produces an ash which has several uses, but is usually used as a nutrient-rich fertiliser. To produce Poultry Litter Ash, the dried litter is burned at temperatures between 800°C and 1,000°C (1,472°F to 1,832°F). This reduces the litter to ash, with any organic material burning, leaving a mineral-rich residue [98].

Burning matter creates polluting chemicals, and burning poultry litter is no different. Incineration releases fine particles of ash, dust, and other materials as particulate matter. Particulates contribute to poor air quality and are harmful to respiratory health when inhaled by humans and animals. Poultry litter is high in nitrogen compounds, some of which convert to ammonia gas during combustion. Nitrogen in manure can lead to the formation of nitrogen oxides during burning. This leads to ground-level ozone, causing further respiratory problems. The oxides dissolve in water vapour and droplets in the air, contributing to acid rain. Sulphur in manure can also create oxides when combusted, having the same effect, creating acid when dissolved in water.

Burning any organic material, such as feathers in poultry litter, produces carbon dioxide. Incinerating large amounts of poultry litter releases large amounts of carbon dioxide. Burning organic matter, especially at high temperatures, gives off **volatile organic compounds (VOCs)** and **dioxins**. Volatile Organic Compounds, a group of carbon-based chemicals, contribute to ground-level ozone and smog. They pose a serious risk to respiratory health and worsen conditions like asthma. Some VOCs, such as benzene and formaldehyde, are also toxic or carcinogenic. Exposure to high levels of these chemicals is a serious health hazard. Dioxins are highly stable compounds that can accumulate in the environment over time. They will bind to organic matter in soil and sediments, remaining in ecosystems for decades or even centuries. They are very toxic and accumulate in the fatty tissue of animals. This results in their concentration increasing up the food chain through **bioaccumulation**. People working on or living close to poultry farms can be exposed to dioxins. They have been linked to developmental and reproductive issues, immune system damage, skin conditions (like chloracne), and endocrine disruption. Dioxins are also classified as known human carcinogens, meaning they can cause cancer. Because they accumulate in fatty tissue, long-term exposure, even at low levels, can be harmful [99].

Culled Chicks

The egg industry uses chickens as egg-producing machines. In nature, chickens lay **10-15 eggs per year**, depending on the breed and conditions. In the egg industry, chickens are manipulated, through intensive breeding and management practices, to lay up to **300 eggs per year** [100]. Also, in the egg industry, males have little worth. In sheds where birds are grown for meat, male and female birds remain together. They don't live long enough to reach sexual maturity, so the sex of birds is not problematic. In egg production, male birds are worse than worthless; they are a problem to be solved.

Organised by sex after hatching, male chicks go one way, females another. Female chicks move to controlled environments, called brooders, immediately after hatching for 6 to 8 weeks. They are then transferred to a growing area to increase weight and size. Called pullets at this point, they live in regulated conditions, ensuring that they reach the right body size and maturity for laying. At 16 to 18 weeks old, pullets move to a laying facility. They will remain in this facility, laying eggs, until they can no longer produce. Being "spent hens," the birds are then culled and used for products like pet food.

> *It's more profitable to grind up billions of baby chicks than it is to raise them for meat*
> *– Seth Millstein* [102]

Due to selective breeding, male chicks at egg-laying facilities are not viable for meat. Considered poor quality meat, they develop little fat and are not commercially viable. They would, though, as living creatures, use resources to survive. It is not worth the investment of food, shelter and water to keep them, so hatcheries kill male chicks straight away [101]. Some hatcheries place chicks in trays or crates in gas chambers that use carbon dioxide or argon. Carbon dioxide use is more common than argon, as although argon gas causes the chick less pain, carbon dioxide is cheaper. Some facilities use more simple methods. Chicks can be suffocated, placed inside a bag that is sealed until the birds expire. Cervical dislocation is a manual method. A worker holds the chick's body in one hand and their head in the other. A sharp tug in opposite directions stretches and breaks the neck, killing the birds. Electrocution, rarely used, kills the birds by electric shock. But the most common method used to dispose of male chicks is maceration. A macerator is a mechanical device with an inlet and output, with fast-spinning blades in the middle. Workers toss male chicks onto a conveyor belt, which carries them over the inlet of the macerator. When they fall in, the blades grind their bodies into pulp. The chicks are conscious during this process.

Ethics aside, looking from a practical perspective, the culling of male chicks produces a lot of waste. Dead birds, culled chicks, and organic waste like egg shells can harm the environment and human health if not managed well. This all adds to the organic waste of poultry farming. Another contributor to this waste is birds that expire before slaughter. Birds often become lame under their own weight as muscle growth outpaces the development of other organs. Their heart, lungs, and even their bones cannot keep the same pace of growth, and broken legs are common. Large-scale chicken meat production hatches billions of chickens each year. Considering that around 4% to 6% of these broilers don't survive to slaughter, we can start to understand the scale of the problem [103]. In 2022, over 70 billion broiler chickens were farmed globally. If 4% to 6% of these chickens die prematurely, this amounts to 2.8 to 4.2 billion birds that don't make it to slaughter each year.

The bodies of these birds will be disposed of, as are the bodies of the culled male chicks. When safe disposal practices, such as composting, are not used, the decomposition harms air, water and soil quality, as well as public health. Carcasses that are disposed of incorrectly or left to rot in open areas release gases such as ammonia, methane, and hydrogen sulphide. The decomposition of bodies will also attract pests, including flies and rodents, which can spread diseases. If dead birds are buried or disposed of near water sources, their decay can lead to contamination of groundwater and surface water. Burial of carcasses, or disposal at landfill, can contaminate soil, overloading it with nitrogen, phosphorus and pathogens.

The Effect of Climate Change

Despite its environmental impacts, poultry is often promoted as a sustainable alternative to red meat. Poultry has a lower carbon footprint per 100g of protein compared to beef or lamb; it also requires less land and water. But this comparison can be misleading when considering the scale of poultry farming and the vast number of birds killed every year. The argument that poultry farming is sustainable is also challenged by its overuse of natural resources and the pollution produced.

> *"Shifting weather patterns, attributable to climate change, are affecting all farmed species. These include high temperatures, rapid and unpredictable temperature fluctuations, high and low rainfall, strong winds, and increased sunlight and humidity. Greater contingency planning may be required to safeguard animal welfare against extreme weather events such as drought or flooding. In addition, physiological changes to farmed species (for example reduced ability to thermoregulate…), may result in climate change having greater impact on their welfare unless such effects are mitigated"* -
> *The Animal Welfare Committee*

Poultry farming requires large quantities of feed, water, and energy. Growing feed crops brings us back to the issues of land and water use. Waste from poultry farms, including manure, wasted feed, and dead bodies, can lead to water and soil pollution. Intensively farming birds releases greenhouse gases, particularly nitrous oxide from manure and carbon dioxide from energy use. Ammonia from poultry manure pollutes the air. This affects the respiratory health of those in nearby communities. It contributes to acid rain, harming plants and reducing soil quality. The clearing of land for feed production destroys natural habitats. It threatens wildlife and reduces biodiversity. The conversion of forests and grasslands to fields of monocrops disrupts ecosystems. Then there is the problem of antibiotic use on poultry farms, used to prevent disease spread in crowded conditions. This leads to antibiotic-resistant bacteria, posing risks to human and animal health. There is nothing sustainable about farming in this way; it is a process that uses resources, destroys some, and replenishes none. Sustainable, when used as an adjective, is: "something that can continue over a period of time or cause little or no environmental damage". This does not describe poultry farming. Poultry farming, like the growing of corn, contributes to and is harmed by climate change.

Although poultry farming contributes to climate change through greenhouse gas emissions, there is also considerable energy use in housing and processing facilities. Running a poultry farm requires significant energy inputs. Ventilation, lighting, temperature regulation, and water purification are not optional.

Poultry houses need constant ventilation. This is to maintain safe air quality, control humidity and prevent the buildup of harmful gases like ammonia from manure. Farmers clean sheds between flocks, not while flocks are in place. In those weeks, manure builds up on the ground, emitting ammonia. This can burn any part of the bird in contact with the ground. Liquid ammonia burns their feet, but when ammonia hovers in the air as fumes, it will burn the eyes of the birds and cause respiratory problems. Farmers mitigate this in part by using mechanised ventilation [104]. This also allows for temperature control, as birds produce a lot of heat as they approach slaughter weight. The heat produced increases as birds become crowded within sheds. In warmer climates or during heatwaves, cooling systems like fans, misters or evaporative cooling pads consume large amounts of electricity. Running these systems, sometimes 24 hours a day, increases the farm's carbon footprint [105].

A cool atmosphere maintains birds' health until they reach slaughter weight. This is better for the birds, but also controls water use. Birds drink about 6% more water for each 1°C rise in temperature from 20°C to 32°C. From 32°C to 38°C, this increases to 5% for each 1°C. They also consume more feed as temperatures rise. Birds consume approximately 1.23% more feed for every 1°C above 20°C. When scaled up, the unit cost multiplied by the number of birds, it becomes cheaper to ventilate and cool a shed than to provide the birds with more food and water. In cooler weather, heating systems keep temperatures optimal, while the use of ventilation continues for air quality, further adding to energy demands.

Lighting is also crucial in poultry farming, as it influences growth, productivity, and egg-laying in hens. Poultry houses use artificial lighting systems which operate 24 hours a day. Lights programmed to maintain the circadian rhythm of the birds use bright lights during the day and dimmed lights at night. This is another system consuming electricity, particularly in closed houses where there is no daylight.

Water is essential for drinking, but also for cooling birds. Many farms rely on purification systems to ensure quality and safety, especially if the water source is from a well or an untreated supply. Filtration and delivery of water use energy, with water purified and pumped across the sheds. These are the systems that allow for the medicating of drinking water at a controlled rate, ensuring the dosing of all birds. Additionally, droughts and water scarcity, exacerbated by climate change, increase water costs and create pressure on water supplies. This pushes farms to use energy-efficient systems to prevent farms from draining natural resources.

Poultry farming is becoming ever more vulnerable to the impacts of climate change. Rising temperatures and extreme weather events, such as droughts, floods and storms, affect feed availability and quality. They make feed less accessible and more expensive for farmers. Heat stress from higher temperatures harms birds, reducing productivity and increasing health issues and mortality. This forces farmers to use more powerful cooling systems, increasing operational costs. Additionally, climate change-induced water scarcity strains the supply needed for drinking water, cleaning and cooling. Climate change is threatening the sustainability and resilience of poultry farms globally. This cycle of impact and vulnerability underscores the need for change within the poultry sector. The same applies to our dietary habits. We need to change to mitigate and withstand the effects of climate change.

The Health Effects of Poultry Farming

The negative health effects resulting from poultry farming are both diverse and complex. The animal agriculture industry has used this argument in court many times. They say that a relationship between poultry farming and health problems does not exist. A relationship does exist, though. There is a clear link between poultry farm waste and illness rates among farm and slaughterhouse workers.

Commercial poultry houses, especially concentrated animal feeding operations (CAFOs), contain high concentrations of airborne particulate matter. The air is contaminated with aerosols composed of organic dust, microorganisms and endotoxins. Even in the open air, these particulates can hover at low levels, where people breathe them in. Exposure to this dust presents a risk for the development of a variety of health problems that can affect the heart and even the brain [106].

Atmospheric particulate matter is a mix of tiny particles, liquid droplets, and microbes. It is classified as respirable (aerodynamic particle size: 0-2.5 μm, $PM_{2.5}$) and inhalable (aerodynamic particle size: 0-10 μm, PM_{10}) for humans and animals. These particulates cause breathing problems, allergies, lung cancer and premature deaths [107]. The respirable mode (< $PM_{2.5}$), can penetrate deeper into the lower respiratory tract. It can enter the lungs and may reach the bronchi and alveoli. By this pathway, particulate matter of this size can access the bloodstream. Studies show that farm workers, particularly those farming poultry, pork and cattle, have higher rates of respiratory illness and disease [77].

Working in an intensive farming environment has several risks. These include physical, chemical and biological harm. The work is demanding. Tasks involve heavy lifting, use of sharp equipment, including knives, long hours and the operation of heavy machinery. There are caustic and dangerous chemicals used for cleaning and other processes. There is also the administration of medicines to animals that are not safe for human use. Biological risks come from exposure to waste, dead bodies, and diseased animals. There are also dust and particulates from feathers, skin, bedding and feed debris. These occupational hazards are unavoidable in these settings. Their properties and concentrations create a combination of health risks and stimulate an inflammatory response not seen in many other industries.

In areas surrounding poultry farms, communities suffer the consequences of factory farming. There are several harmful gases emitted in animal agricultural settings. These include **methane**, **ammonia**, **hydrogen sulphide**, **carbon monoxide** and **carbon dioxide**. These pose serious health risks for both workers and animals. Additionally, many volatile organic compounds in the CAFO environment produce unpleasant odours. These odours create poor working environments. They also create poor health in the communities around CAFOs. One of these is that smells can become better or worse due to changes in wind direction. This leaves people feeling unable to open windows, even on the hottest of days. It also lowers the value of properties near farms.

It is routine for poultry farms to deal with bird mortality due to disease, poor conditions, and the culling of male chicks in egg production. The bodies are often burned, causing smoke and the smell of burning bodies to pollute surrounding areas. Bodies may be buried or dumped in pits or lagoons, depending on local regulations. Decomposing carcasses harbour harmful bacteria like Salmonella, E. coli and Campylobacter. These germs can infect humans through contaminated air, water or soil [108]. The pathogens are extremely dangerous, especially for the very old and the very young. They can cause foodborne illnesses, gastrointestinal infections and other serious health issues. Dead birds, particularly those that die from infections or in unsanitary conditions, may carry diseases, like avian influenza (bird flu). These diseases have the potential to spread to humans from birds or through contaminated surfaces, air or water.

CAFOs store many of their contaminants on-site. Animals produce waste at a constant rate. For easy management, it is often stored in open-air sheds, close to the animals. This is in addition to the pits and lagoons, creating another source of pollution and health risks, as bacteria breed in such situations. These waste stores have the potential to cause a devastating disease outbreak.

Endotoxins, components of bacterial cell walls, can cause a strong immune response, resulting in inflammation of the lungs or respiratory distress. It can also worsen asthma or other pre-existing respiratory conditions. Long-term exposure to endotoxins can harm the lungs, increasing the chance of developing chronic respiratory diseases. Endotoxins build up in high concentrations in manure, dust, and other organic materials. When manure or bedding materials break down, they become airborne, making them impossible to avoid.

Full-time barn workers, veterinarians and locals face exposure to several noxious substances. Common symptoms include irritation of mucous membranes and eyes, nasal congestion and a runny nose. There are also complaints of wheezing, coughing and dyspnoea (shortness of breath). People will develop asthma, asthma-like symptoms, worsening of pre-existing asthma, chest tightness and exercise intolerance. It is particularly common for barn workers to experience measurable annual declines in their lung function.

Case Study – Bangladesh

In Bangladesh, poultry farming is a traditional practice, with people keeping small numbers of birds at home, often in a backyard [109]. Waste from the birds was collected in bags and used as fertiliser for crops that people would grow for themselves. As this process has become more industrialised, the commercial poultry sector has grown at a steady pace of 15-20% per year. The waste, increasing at the same rate, has become more of a problem. This has resulted in poor air and soil quality and environmental deterioration. People are also suffering from deteriorating health and significant amounts of greenhouse gases. Poor waste management has also caused health and welfare problems in flocks. There are bad smells and flies breeding among the birds.

Poultry waste can consist of a combination of the debris of farming birds. Manure, feathers, spilled feed, dead birds and broken eggs can all be in the mix. There is also wastewater, litter or bedding materials and slaughterhouse waste. A flock of laying birds, consisting of 3,000 birds, can produce 300 kg of faeces per day. They will consume 500 kg of feed. This results in the generation of waste at a rate of 0.1 kg/bird/day. Most of this poultry manure and litter will be applied to agricultural land. The overuse of manure is causing surface and groundwater to become saturated with nitrogen and phosphorus. The contamination is causing a drop in crop yields, as fields are being poisoned. Composting is possible when waste isn't used as fertiliser. This process requires more land, financial investment and labour, yet still emits harmful CH_4 gas, as manure decomposes. Hatchery waste is still sent to landfill, creating another source of soil, water and air pollution.

In Bangladesh, poultry litter harms air quality and the environment due to high ammonia levels. The poultry waste is contributing to global climate change by emitting greenhouse gases. These include nitrous oxide, CO_2, and CH_4. The air pollutants released in the farming of birds have a significant negative impact on both human and animal health. Long-term exposure to sustained air pollution leads to allergic reactions and reduces life expectancy. There is also the risk of human and animal infections, such as zoonotic avian influenza. Food or water contaminated by poultry waste may contain pathogens and water pollutants. These cause gastrointestinal diseases like typhoid fever, cholera and hepatitis E infections. It is common practice to discharge poultry waste into water systems using channels. These channels drain into ponds and rivers, contaminating them. This could lead to surface and drinking water containing heavy metals, antibiotic residue, and microbes.

Polluting the Pocomoke River of Maryland (2012)

In 2013, the Waterkeeper Alliance Inc. took civil action against Alan Hudson, a supplier of beef and poultry products to Perdue Farms [110]. The Hudson family farm kept 40,000 chickens at any one time and used natural waterways for waste disposal. Court filings stated that the farm was responsible for high levels of pollutants in a tributary of the Pocomoke River.

The proceedings revealed that "alarmingly high levels of faecal coliform, E. coli, nitrogen and phosphorus had been discharged from Hudson's farm and that at least some of those contaminants would reach the Pocomoke River". But, because the Waterkeeper Alliance argued that the pollution was a result of poultry waste, the judge denied the motion. There was no argument that the Hudson farm was, at least partly, responsible for the river's pollution. Contamination levels were high enough to be a health hazard. But the Waterkeeper Alliance had complained about poultry farming and not beef farming. The complaint stated that chicken manure and bedding waste were entering the river from the Hudson farm. Aerial photographs of the farm were said to show a large uncovered mound of poultry manure near the chicken houses. Trenches in the ground around the pile channelled runoff to a nearby drainage ditch. This ditch was part of a series of ditches on the farm used to channel rainwater and waste. It had been dug to drain into the Franklin Branch of the Pocomoke River. An inspection found that the mound in the photos was not poultry waste. It was fertiliser bought by Hudson for a field. David Bramble, a district manager from Maryland's Department of the Environment Water Management Administration, inspected the Hudson farm. He reported that "the farmer had approximately 42 head of beef cows, that are being fed in a small dirt field, with manure coming into contact with stormwater, which was contributing to the farm runoff to the open ditch."

During the relevant time, October 2009 to April 2010, the Hudson Farm had between 85 and 90 cows and calves, 40 brood cows and calves and one bull. Each cow produces roughly 27-36 kg of manure per day. This means that the cattle on the Hudson Farm produced approximately 1361 kg of manure daily. A considerable amount of this manure was being channelled into the river.

Although the manure contamination of the river was a clear finding in the case, no action resulted against Alan Hudson. The Maryland Department of Environment sought a fine of $4000 for improper handling of animal waste. But this was denied, and the pollution of the river continued.

A study – The Negative Impacts of Poultry Farming

In February 2023, the journal Science of the Total Environment published a study investigating the impact poultry farming has on human health and the environment [77]. The study, *Intensive poultry farming: A review of the impact on the environment and human health*, Goran Gržinić et al., analysed data collected from other papers on poultry farming. The data documented the poultry process, the waste produced, and the negative impacts of the practice.

The study explored the negative impacts that different aspects of poultry farming had on the air, soil and water. They then compared how these harmed humans who came into contact with the pollutants. The waste by-products of poultry farming include NH_3, N_2O and CH_4 emissions. These all contribute to global greenhouse gas emissions and damage animal and human health. Pesticide residues, microorganisms, pathogens, and pharmaceuticals like antibiotics and hormones contribute to pollution. Improper ratios of metals and macronutrients in manure also worsen the problem. They can also lead to the formation of antimicrobial/multidrug-resistant strains of pathogens.

The dust emitted from intensive poultry farming consists of feather and skin fragments, faeces, feed particles, and microorganisms. It damages the health of farm workers and those living close to these businesses. Another aspect degrading the lives of those living near poultry farms is the smell. The odours emitted from these farms can ruin the quality of life and health of workers and residents. Ventilation in poultry barns, meant to be for the birds, expels pollutants inside the barn to the air outside. Significant contamination of the area within the confines of the farm can result. Also, the wind, able to carry these pollutants for miles, means that the foul air is a hazard to everyone nearby and some not so nearby. The smell that comes from poultry farms is hard to imagine for the average person. It is pungent. It hangs in the air and can have a feeling of coating the skin.

> *Poultry farms generate fastidious odours containing dimethylamine, ammonia, ketones, aldehydes, organic acids and other compounds which can have adverse effects on the quality of life and health of farm workers and surrounding population - Carey et al., 2004; Nowak et al., 2016*

Noise pollution from poultry farming, or farming in general, is seldom considered. Most of the noise on farms comes from the animals, with some being louder than others. Sound levels inside a poultry barn will range from 50 dB to 90 dB during the daytime when birds are awake. When farming equipment is in use, noise levels can exceed 90 dB, a level known to cause fear and stress in birds. The recommended exposure limit for noise is 85 - 90 dB for workers, depending on the country. This means sound can be at the upper limit of allowable noise for large parts of their working day. Outside the barns, at 15-20 m, noise levels will measure between a range of 44 to 63 dB.

Greenhouse gases are a by-product of poultry farming, with manure being the greatest contributor. The levels of greenhouse gases will change depending on how farmers handle, store and use them. Nitrous oxide releases are, for the greater part, from storage and the spreading of litter on fields. It happens through the nitrification and denitrification process - $NO3 \rightarrow NO2 \rightarrow NO \rightarrow N2O \rightarrow N2$.

Intensive poultry farms are a known and obvious source of biohazards. People use compulsory PPE when entering poultry barns, including wearing face masks, boot covers and overalls. Disinfectants, used when leaving, prevent the spread of biohazardous material and disease. Within poultry farms, whether ducks, chickens or other birds, it is common to find organisms that feed on decaying organic waste. These organisms include fungi and pathogenic microorganisms. They have the potential to become airborne and blow outdoors. Studies conducted on the potential of these microbiota have found they can travel up to 3000 m from poultry barns.

The 2023 study also found that the use of manure was a consistent source of nitrogen and phosphorus pollution in water. Organic matter, PO_4^{3+} and bacterial microorganisms were the main pollutants from poultry farming. When compared to other forms of animal agriculture, the water quality nearest poultry farms is the lowest. Pharmaceuticals, excreted by animals in urine and faeces, can also contaminate water. They disperse through improper storage or use of manure as a fertiliser. Testing has detected these medications in water sources. From surface water to drinking water, their presence has several negative impacts. These include the disruption of marine life's endocrine systems and the development of antibiotic-resistant bacteria.

Medications, whether given to humans or animals, are not completely metabolised. Roughly 30-90% is excreted completely unchanged. In tests of poultry farm wastewater, more than half of the samples, 55.6%, had detectable antibiotic residue. The use of antibiotics in farming has caused water ecosystems to become significant reservoirs of pharmaceuticals and antibiotic **resistance genes (ARGs)**. These are specific genes in bacteria that provide them with resistance to antibiotics. The genes enable bacteria to survive, grow, and even multiply in the presence of antibiotics. In normal circumstances, these antibiotics would kill them or stop their growth. Once bacteria carry ARGs, they can block specific antibiotics using enzymes that break down or change the antibiotic. They can alter the antibiotic binding site, expel antibiotics from bacterial cells, or stop antibiotics from entering.

Soil quality is damaged by the same contaminants as water sources. Phosphorus, nitrogen, microbes, antibiotics and other medications poison the ground. A study by Parente et al. (2021) in Brazil looked at antibiotic contamination in soil [111]. It showed that antibiotics accumulate in soil with regular use of poultry litter as a fertiliser. The build-up of antibiotics in soil, from manure or litter, increases the number of antibiotic-resistant genes in bacteria. The presence of antibiotics also harms invertebrates living in soil. Earthworms living in soil polluted with antibiotics show changes in behaviour. They also have low reproduction rates and increased mortality. Earthworms play a crucial role in maintaining healthy ecosystems. Through their activities, they benefit soil health, plant growth and the greater environment. The recycling of organic matter is reduced without earthworms. Soil quality reduces with poor soil structure and nutrient levels. Earthworms carry organic material deeper into the soil, contributing to soil carbon storage. Earthworms and other invertebrates that live on or under the ground are vital to healthy soil. They support plant growth and provide balance and structure to the ground.

The large amounts of phosphorus that accumulate in the soil, through the overuse of manure, also reduce the iron and zinc content of plants. This impacts humans, as the iron and zinc levels in fruits and vegetables are less reliable. This has been linked to symptoms of deficiency of these two nutrients in people. In plants, zinc deficiency inhibits plant growth, causes yellowed leaves, and can even cause necrosis.

Another significant problem caused by poultry farming is the bird flu virus. Bird flu creates a financial burden for farmers and a public health risk. Bird flu, also known as avian influenza, negatively impacts poultry farming in several ways. These range from economic losses to long-term operational challenges. Bird flu, especially **highly pathogenic avian influenza**, causes severe illness and death in poultry flocks. Affected farms can lose entire flocks within days; this can mean a loss of tens of thousands of birds in one go. Once detected, bird flu requires mass culls of birds. This is necessary to prevent the spread of the disease and possible human transmission. The death of so many birds, from illness and culls, is expensive. Farmers must pay for the killing and then disposal of the bodies of birds. Farmers also lose money from not selling birds or eggs. Once all birds are removed, barns, equipment and farms are cleaned. All cleaning must be thorough to prevent a repeat outbreak. Without these precautions, trade restrictions and export bans can hit farmers hard. Reduced demand also leads to financial losses. This situation can harm the economies of countries that depend on poultry exports.

The natural reservoir of the bird flu virus is burrowing poultry, including chickens and turkeys. Wild birds can also carry the virus over long distances. The study from Goran Gržinić et al. continues with the outbreak in 1997 of an avian flu strain that was able to infect humans. Once only found in birds, this flu strain is now found in humans across the world. According to the World Health Organization (WHO), there were 862 such cases worldwide by the end of 2020. Of these cases, 455 (52.8 %) resulted in death. Inhaling infectious droplets or dried debris or direct contact with sick birds can cause infection. The result is severe influenza and pneumonia, with the potential to escalate to respiratory failure.

Bacteria within poultry farms are a significant health risk for those who work within them. Levels of bacteria in most poultry farms exceed the threshold limit value for this type of working environment. Bacteria and the endotoxins that they produce can cause respiratory disorders and diseases. They can also cause fever, disturbances in gas exchange, bronchospasm or allergic reactions. Exposure to the bacteria found in poultry barns can be serious. Purulent infections, poisoning, and, in extreme cases, sepsis are all possible. Outside of the poultry barns and outside of farms, the transfer of antibiotic-resistant bacteria may occur. This can result from direct human contact with farmed birds or the spread of bacterial strains in outdoor environments. This includes the use of poultry manure and litter as fertiliser. German studies have investigated regions dense with farms, with more than 12 farms per 500 m area. People in these regions have a lower quality of life. Also, they have higher rates of lung-related health issues. The concentration of people suffering from such issues increases the closer they live to the farms. Rates of pneumonia, eye and nose irritation and gastrointestinal illnesses are higher in those living close to poultry CAFOs. The Netherlands has similar outcomes, with higher rates of pneumonia in communities close to poultry farms [112].

Antibiotics are present at every stage of poultry production, from chick to chicken, duckling to duck, gosling to goose. This use has resulted in antibiotics being detectable in meat samples from slaughterhouses. They are also present in packaged meat and ready-to-eat products sold to consumers. Testing has shown that antibiotic levels in meat can be higher than allowed maximums. The results of these tests have also found that, in more than 50% of cases, more than one medication is present. Eating poultry can mean consuming penicillin, as it is used on a regular basis in poultry farming. Consuming this meat creates a risk of developing penicillin allergy or triggering a reaction in those sensitive to the drug. Cooking does not make the meat safe, as heat treatment does not destroy antibiotic residues. They are completely unavoidable if one eats these meat products. Long-term exposure to antibiotics through meat may have carcinogenic and teratogenic effects. This means they may cause cancer and birth defects. These antibiotics may also damage fertility and induce antibiotic resistance. Chloramphenicol is an antibiotic that is banned from animal farming in the USA, Canada, Australia, and the European Union. It is a known carcinogen, but is still used in developing countries as a growth stimulant.

Another risk from eating meat with residual antibiotics is the effect of these compounds on the host's microbiome. Our microbiome consists of roughly 7000 different species of bacteria. 95% of these strains are probiotic and commensal bacteria. 5% are opportunistic bacteria, bacteria that lead to infection and illness. Constant exposure to antibiotics leads to disturbances in the composition of our microbiome. It also allows the overgrowth of opportunistic bacteria. Overgrowth of opportunistic bacteria causes inflammation and the development of colorectal cancers. A more insidious danger is that babies are at risk of ingesting antibiotics through their mother's milk and first solid foods. Dinleyici et al. (2018) found β-lactam antibiotics in the milk of lactating women who did not take antibiotics during their pregnancy. In children, these antibiotics may contribute to immune system disorders, obesity and abnormal bone development, including the teeth [113].

Antibiotic resistance, from overuse in meat production, is the greatest threat to human health, as stated by the WHO in 2021. In the EU, 33,000 people die each year from untreatable antibiotic-resistant bacterial infections. Bacteria resistant to the "last-resort antibiotics" cause 30% of these deaths.

Huynh v. Blanchard

In July 2021, a Texan judge issued a permanent injunction against a poultry farm. In this case, the jury determined that the smell from the farm caused a nuisance and devalued the properties in the surrounding community.

In 2015, Steve Huynh purchased 231 acres of land in Malakoff. He intended to use the land as a chicken farm for Sanderson Farms, for whom he had owned and operated chicken barns since 2002.

Knowing that there would be complaints from locals, Sanderson approved the property as a barn site [114].

Soon after the chicken barns began to operate, people living nearby noticed a pungent odour coming from the barns. In the lawsuit, the Appellees complained about the smell. They claimed that the smell from the chicken barns stopped them from enjoying their homes or going outdoors. They complained many times to the Texas Commission on Environmental Quality (TCEQ). The TCEQ investigated the complaints and issued notices of violation, which Huynh ignored. The smell continued to invade the lives of the community. Unable to cope anymore, residential groups filed lawsuits against Huynh. Their complaints were fraud, nuisance, trespass and intentional interference with property rights. The lawsuit asked for financial compensation for the loss of property value, but the judge rejected this. The judgment was a permanent injunction, preventing any more poultry farming at the site.

Having purchased the land in 2015, Steve Huynh had constructed two farms. These allowed him and those he worked with to have sixteen barns and grow 444,800 birds per flock. This number of birds is twice the number considered "likely to cause a persistent nuisance odour" under the TCEQ guidelines. The two farms were also very close together, separated by only 91 m.

Sanderson Farms delivered the first flock of chickens in June 2016, and by November 2016, the barns were in full operation. The chickens produced approximately 4.5 million kg of manure each year, with each flock having an approximate mortality rate of 5%. They discarded the bodies of dead chickens in composting sheds in layers and covered them in wet litter, saturated with manure. Sanderson's division manager testified, via deposition, that dead chickens, like most dead animals, have a rotting odour. Dr. Albert Heber, the Appellants' expert witness, testified that chicken manure is "offensive" and "smells bad." The manure also generated ammonia and hydrogen sulphide, adding to the community's distress.

Residents raised complaints, and the farm was found to violate two different Texas codes. Both prohibit the release of chemicals, contaminants or pollution that cause suffering, or stop people from enjoying their property. Residents could no longer use outdoor spaces. Activities like horse riding and outdoor entertainment stopped. The smell would even get into people's homes. One resident, Mersini Blanchard, testified that it smelled like "chicken poop and dead animals all together." She stated that the smell made her gag and "sick in [her] stomach" on more than one occasion. Due to the suffering the farm caused, the judge issued a permanent injunction, stating that all farming activities were to stop.

Veganism as a Solution

Adopting a vegan diet isn't only good for the farmed birds, the ducks, geese and chickens. It would end the environmental impacts associated with poultry farming. Removing the demand for birds produced as commodities would reduce pollution, resource use and greenhouse gas emissions.

It is inherent that plant-based diets are more sustainable. They require fewer resources and generate less pollution. Producing plant-based protein like beans, lentils, and tofu uses less land, water and energy than raising poultry flocks. These foods also cause less nutrient runoff, no pathogen contamination and no antibiotic resistance. The best argument for not farming poultry can be found in regions where it has already stopped. Areas that were once damaged by poultry farming, but are now restored and thriving.

Chesapeake Bay, USA

The Chesapeake Bay watershed has seen serious environmental issues. A concentration of poultry farms on the Delmarva Peninsula is well known for its runoff. The Chesapeake Bay region rears more than 1 billion birds for human use every year [115]. The Chesapeake Bay Restoration Project, started 40 years ago, has reached its critical 2025 deadline. In 2010, the group set a target for the greatest pollution levels. This was to reduce eutrophication caused by pollution from farms. The harmful algae blooms degraded water quality and harmed aquatic ecosystems [116]. Dead Zones, along with the algae blooms, have devastated the region. Fish and shellfish numbers have been declining with increased pollution. Stricter regulations on poultry farm waste management have reduced some runoff. There has also been a shift away from intensive poultry farming in some areas. Regulations reduced poultry farm density by doubling the smallest allowed distance between farms. Fewer farms mean fewer pollutants.

A direct action was the shutdown of the Eastern Shore chicken rendering plant in December 2021. Following complaints, the Maryland Department of the Environment mandated that the Valley Proteins Inc. plant in Linkwood pause operations in December 2021 until it could meet wastewater discharge limits. they also had to reduce the risk of overflows from their storage lagoons. If the obligations weren't met, the plant would be fined. If the discharges continued, the plant's permit was going to be suspended [117].

The shutdown order came after a series of MDE inspections found several issues. There was an illegal discharge into a holding pond, discharges of sludge, leaks and overflows from treatment tanks. Wastewater, not made safe, was being released into a stream that led to the Transquaking River. The inspections also found ongoing acidic wastewater releases into a stream. Chlorine-treated wastewater was leaking onto the ground. Foam and wastewater overflowed from a treatment tank, and raw chicken waste was dumped on the ground. None of this had gone unnoticed; both locals and environmental groups had complained for years about the Valley Proteins plant.

The plant produced pet food by rendering up to 1.8 million kg of chicken entrails and feathers from poultry processing plants. Much of the waste from processes was discharged into the Transquaking.

This river flows into Fishing Bay, a Chesapeake Bay tributary and has been harming the water with nutrient pollution for more than 20 years. The rendering plant was the river's largest single source of pollution of this type. The water suffered from algae blooms resulting from the rendering plant and its waste.

Actions, like closing the worst polluters and decreasing the number of chicken farms, have improved water quality. Levels of nutrient pollution are declining, and the ecosystem is recovering. These improvements have benefited fish populations. Shellfish, like oysters, have started to return, and the biodiversity of the bay has improved. Healthier water quality has also affected local communities that rely on the bay. Fishing is now possible, and tourism has boosted the local economy. There is still a long way to go. Chesapeake Bay still faces the threat of ongoing, industrial-scale pollution from poultry farming. But with community groups fighting for the land, water and air, there is hope.

Poultry Farm Closures in North Carolina, USA

North Carolina is the top producer of eggs and poultry in the United States. But, in recent years, the number of farms in North Carolina has decreased. Meanwhile, the amount of land used has remained steady, as farms have increased in size. The United States Department of Agriculture National Agricultural Statistics Service states that 15,640,000 birds laid 4,075 million eggs in 2020. North Carolina produced 961,300,000 broiler birds and 30,000,000 turkeys; this is big business [118]. The environmental impacts of farming so many birds are well-documented.

The names of poultry farms, closed for environmental concerns or regulatory action, are rarely released to the public. The closures often involve larger industrial farms or concentrated poultry operations that were targeted for their negative environmental impacts. The closure of poultry farms in North Carolina has, for the most part, been due to voluntary actions by farm operators. It is easier for them to move than deal with regulatory pressures and public policy shifts.

North Carolina concentrates most poultry farms in the eastern part of the state. The industry has been a major economic driver. But these farms are facing increasing pressure to clean up their practices. Those that don't face fines, suspensions, or even closure. Regulations were also changed to protect the state's rivers, such as the Neuse and the Cape Fear. These rivers suffered from the release of excess nutrients. Nitrogen and phosphorus leaching into the water from poultry waste polluted the water. The excess nutrient load was causing eutrophication, predictable algal blooms and hypoxia. The closure of these poultry farms resulted in better water quality and reduced air pollution. Water bodies began to recover, and algae blooms decreased. This all improved the water's oxygen levels, allowing aquatic wildlife to return. Improved air quality, including foul smells, allowed people to enjoy outdoor spaces again.

North Carolina aims to lessen the impact of industrial poultry farms. State initiatives, government agencies, and environmental groups back this effort. The North Carolina Department of Environmental Quality (NC DEQ) has strengthened regulations on waste management. They also started to enforce stricter pollution controls. Additionally, in 2018, North Carolina passed laws for more sustainable waste management on large-scale animal farms.

Rewilding The Netherlands and Beyond

The Netherlands had a severe nitrogen pollution crisis. A major cause was high-density livestock farming, including poultry.

The Netherlands has one of the world's most intensive farming systems. This results in very high levels of ammonia emissions from livestock manure. The ammonia reacts with nitrogen oxides from transport and industry. This combination has created a surplus of reactive nitrogen in the environment.

Ammonia from poultry manure, contributing to fine particulate matter ($PM_{2.5}$), harmed human health. Excess nitrogen has caused soil acidification. The harm to plant life has reduced biodiversity in heathlands, forests and wetlands. Runoff from fertilisers and manure produced high nitrogen levels in rivers, lakes and groundwater. The result was eutrophication that killed aquatic life. Many natural systems were being endangered. The pollution ruined many sensitive ecosystems like heathlands, which rely on nutrient-poor soils. Nitrogen-loving plants (e.g., grasses and weeds) have outcompeted native species, reducing biodiversity. Species, including butterflies, amphibians, and birds, have declined as the whole ecosystem shifted.

Pollution levels in the Netherlands violated EU environmental laws. This included the Nitrates Directive and the Habitats Directive, which mandate the protection of water quality and biodiversity [119]. The EU pressured the Dutch government to take urgent action. The Dutch Council of State ruled that the government's nitrogen reduction policies were inadequate. In response, the government stopped construction projects. They applied stricter scrutiny of agricultural emissions and made changes to farming practices to reduce emissions. Unhappy, farmers protested the restrictions. But environmental groups stressed the urgent need for action to protect ecosystems and public health.

To reduce nitrogen emissions, the government offered voluntary buyouts for farmers. They targeted those near protected areas [120]. Livestock farming operations stopped on the farms that were sold, reducing nitrogen output. This was done as part of a huge rewilding project. Activities included re-flooding former farmlands to recreate natural wetlands. Water quality also improved, supporting biodiversity. They restored heathlands and forest lands to their original state and reintroduced native plants as part of the project, encouraging wildlife to return. The rewilding also created connected habitats for wild species to migrate through. The goal is to cut nitrogen emissions by 50% by 2030, meet EU nitrogen targets and restore balance to ecosystems. The rewilding projects are a cornerstone of this strategy.

The successes of the rewilding have included:

- Wetlands and heathlands that can support species like lapwings, storks, and otters.
- Natural vegetation has been restored, and eutrophication is reduced.
- Restored peatlands and forests now act as carbon sinks, aiding climate change mitigation.
- Rewilded areas are attracting visitors, tourists and school groups, boosting the local economy and raising environmental awareness.

Phasing Out Poultry Farms in South Korea Following Avian Influenza

South Korea has faced significant challenges in managing poultry farming. Particular problems were sales of contaminated eggs and outbreaks of Highly Pathogenic Avian Influenza (HPAI).

Between 2017 and 2018, South Korea faced egg contamination scandals. Contaminated eggs contained fipronil, an insecticide used to kill fleas and ticks. The U.S. EPA classifies this chemical as a "possible human carcinogen". Eggs were also contaminated with bifenthrin and DDT, a toxic pesticide banned in the 1970s [121].

The contamination came from the overuse of pesticides in crowded poultry farms. Farmers resorted to banned chemicals to address infestations that they could not control. The cramped conditions in battery cages used in egg production made the infestations worse.

Between 2014 and 2024, large outbreaks of HPAI led to the culling of tens of millions of birds. This disrupted poultry production and raised public health concerns around farming practices. To stop the spread of illness, some high-risk farms closed. Others shifted to more eco-friendly methods, such as reducing the number of battery cages and some places banned live poultry sales in markets [122].

When poultry farms closed or production was reduced, there was a clear reduction in the transmission of avian influenza. Avian flu can have serious consequences for both human populations and poultry farmers. The removal of poultry farming in some areas reduced the risk of further outbreaks. The decline in waste and pollutants from abandoned poultry farms allowed water quality to recover. This created better conditions for wildlife, restored ecosystems and community areas.

Conclusion

Poultry farming, thought of as more sustainable than other forms of meat production, is terrible for the environment. Nutrient pollution, greenhouse gas emissions, resource use and antibiotic resistance all contribute to ecological damage. Damage that gets overlooked. We should also not ignore that those who live near these farms suffer. They can no longer enjoy their gardens or outdoor spaces. Areas such as parks or fields are ruined by disgusting smells. These smells also invade their homes and damage their health.

People who work in these farms often cannot change jobs due to few other industries operating in the area. So, they must endure the many health problems caused by their work. Burning eyes, skin irritation, asthma and other respiratory illnesses are normal within these workplaces.

Removing poultry from your diet, or choosing a vegan diet, is a powerful way to mitigate this damage. Choosing plant-based foods reduces the demand for poultry and supports sustainable and healthy farming. People believe that one person makes no difference. They use this reason to not sign petitions, not protest, not vote. But when you vote with your money, companies listen. Consumer demand is a powerful driver of change in the food industry. This is also true with the rise of veganism and the increase in meat and dairy-free products. The global plant-based food market is expecting significant growth.

Reaching an estimated $95.52 billion by 2029, with double-digit annual growth rates. Companies are recognising this as an opportunity to cash in. There are examples of this change. Burger King, McDonald's and KFC have introduced plant-based options. This is only in response to the demand for meat-free fast food. Dairy-free options include brands like Oatly and Silk. Both companies have expanded their product lines to offer more dairy-free alternatives.

In the following chapters, we'll continue to explore the environmental impacts of animal farming. As we do, it will become clear that a shift towards veganism is key for the health of our planet. Animal agriculture needs vast amounts of soya, corn and grain to continue. Transitioning to plant-based diets reduces this extreme production. It helps promote more sustainable practices, such as organic farming, permaculture and agroecology. Practices that prioritise soil health, biodiversity, and water conservation.

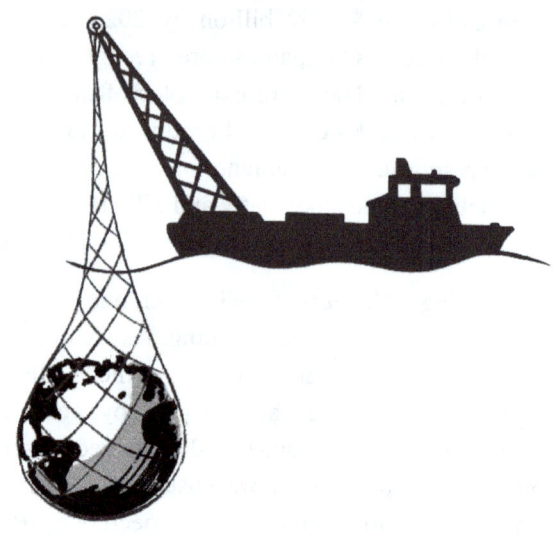

Chapter 5: Fishing and Farming the Oceans

The Beginning of Fishing

Humans catching fish for food dates back to ancient times, beginning in the Upper Palaeolithic period, around 40,000 years ago. Human remains from this period show that people regularly ate freshwater fish. Items like mollusc shells, fish bones and cave paintings prove that seafood was important to those living near seas and oceans.

Across the world, humans took up fishing at different times throughout history. Developing different techniques and equipment to catch marine creatures. Corralling fish, catching them by hand, spearfishing with barbed poles, and casting nets were all used. The belief was that fish would always replenish themselves. There were so many fish in the waters that the supply was endless. But fishing techniques improved and became commercialised. This meant people could catch fish faster than fish could reproduce.

Overfishing

The oceans cover over 70% of the Earth's surface and are vital to the health of our planet. They regulate the climate, balance temperatures, and absorb carbon from the atmosphere. They also provide food for countless species, supporting both marine and terrestrial life. Yet today, the fishing industry poses one of the greatest threats to aquatic ecosystems. Overfishing, destructive fishing practices, and bycatch have led to the degradation of marine habitats. Fishing is disrupting ecosystems crucial for maintaining the balance of life on Earth.

> *Atlantic salmon is an indicator species, reflecting the health and cleanliness of marine and freshwater ecosystems. A shrinking salmon population is a warning sign that much more work is needed to improve our natural environment. - Environment Agency and Natural England*
> (123)

Overfishing is removing fish and other marine creatures faster than they can reproduce. This depletes fish populations and disrupts aquatic ecosystems. It often results from unsustainable fishing methods, such as trawling and the use of large nets. These techniques catch immature animals and non-target species (**bycatch**) while damaging ocean habitats. Overfishing impacts whole ecosystems. It disrupts food webs, causing ripple effects throughout marine environments. Predatory species may decline due to a lack of prey, while other species can overpopulate. This causes imbalances that affect coral reefs, seabed habitats, and even water quality. These changes can harm ocean ecosystems. They make them more sensitive to other stresses like climate change and pollution. The loss of biodiversity is one significant consequence. But so is the damage done to the vital ecological processes that support life, both in the ocean and on land.

Overfishing has caused the collapse of many fish populations around the world. The number of fish, such as tuna, cod and haddock, has dwindled. Since the 1970s, the number of wild salmon has fallen by more than half. According to the Food and Agriculture Organization (FAO), approximately one-third of global fish stocks are overfished. Fishers are catching over 60% of fish at the limit of their sustainable levels, meaning fish populations are at risk of collapse [124].

Sustained overfishing will cause critical declines in breeding populations. There will be so few adult and juvenile fish remaining that they can no longer sustain themselves. The overfishing of some animals, such as sharks, has disrupted entire ecosystems. A fishery's ability to recover depends on conditions remaining positive for restoration. Significant shifts in species combinations can lead to a new ecosystem equilibrium. Changes to the animals in the habitat alter energy flows and species dominance. For instance, overfishing of trout means carp may take advantage of the reduced competition. They establish dominance and prevent the trout from re-establishing a breeding population.

A 2022 report from the Food and Agricultural Organization stated that "the percentage of stocks fished at biologically unsustainable levels has been increasing since the late 1970s, from 10 per cent in 1974 to 35.4 per cent in 2019." Put another way, 35.4% of the world's fish species are under threat. This comes as a direct result of people wanting to eat them.

Techniques and Bycatch

After World War II, there was significant growth in global fishing activities. Intensive fishing operations spread to cover most fisheries. Techniques in commercial fishing include:

- **Trawling:** Trawling drags a net along the sea floor (bottom trawling) or through a water column (midwater trawling). The process catches fish and seafood that live at different depths in the water.
- **Longline Fishing:** Longline fishing uses a long line with thousands of baited hooks, laid out on the water. It is used to catch species such as tuna, swordfish, and halibut.
- **Purse Seining:** Purse Seining sets a net to encircle schools of fish and draws the bottom closed like a purse. This is used to trap species such as sardines, anchovies, and mackerel.
- **Gillnet Fishing:** Gillnetting uses vertical nets set in water, designed to entangle fish, such as salmon or cod, by their gills.
- **Trap Fishing:** Trap Fishing uses baited traps or pots. They are laid on the sea floor to catch species such as lobsters, crabs, and some bottom-dwelling fish.
- **Dredging:** Dredging drags a heavy frame with a mesh bag across the seafloor to collect shellfish such as scallops, clams, and oysters.
- **Drift Net Fishing:** Drift Netting catches fish by using large free-floating nets to capture anything that is swimming in open water. This method is controversial due to **high bycatch rates**, but it is still used.
- **Pole and Line Fishing:** Pole and Line Fishing, a more traditional fishing method, is still used to bait and catch fish such as tuna.
- **Fish Aggregating:** Fish Aggregating Devices (FADs) are floating devices that attract fish. Gathering the animals

together in the water making them easier to catch. These are commonly used in tuna fisheries but, like other techniques, lead to bycatch.

Fishing methods that scrape the sea or ocean floor, bottom trawling and dredging, are the most controversial. These techniques are devastating to coral, sponges and other slower-growing species. Unable to recover, these creatures and their habitats can be forever lost. This alters how the ecosystem works and can permanently change the biodiversity of the environment.

Bycatch causes the wholesale capture of incidental species. It is estimated that 7.3 million tonnes of non-target catch (bycatch) are discarded every year. It accounts for over a quarter of all marine creatures caught. These creatures are returned to the ocean, only to die from injuries or exposure. Having been out of the water for some time, they cannot survive. In the case of shrimp, the mass of bycatch is five times larger than that of the shrimp caught in temperate areas. It can be up to ten times greater in tropical regions [125].

> *It is estimated that over 300,000 small whales, dolphins and porpoises die from entanglement in fishing nets each year, making this the single largest cause of mortality for small cetaceans. Species, such as the vaquita from the Gulf of California and Maui's dolphin from New Zealand, face extinction, if the threat of unselective fishing gear is not eliminated. – WWF* [126]

Consequences of Bycatch

Bycatch causes the unnecessary death of millions of marine animals each year. Dolphins, whales, sea turtles, and even seabirds are often caught in fishing gear. Many of these animals are already threatened or on endangered lists. The added pressure from bycatch is pushing them ever closer to extinction. For example, the vaquita, a small porpoise found in the Gulf of California, is critically endangered. Fewer than 20 individuals remained in the wild as of 2018. The last count in the summer of 2023 reduced the number to 10-13. The primary threat to the vaquita is becoming bycatch in gillnets used to catch another endangered species, the totoaba fish. Despite efforts to ban these nets, illegal fishing continues, driving the vaquita towards extinction [127].

Authorities banned gillnet fishing in the Upper Gulf of California over seven years ago, but this did not deter fishers, and they continued hanging nets within the protected area. To prevent this, the Mexican Navy stepped in. In 2022, they placed concrete blocks set with rebar hooks in a zone called the Zero-Tolerance Area (ZTA). In November 2023, they sank more blocks outside the protected area, increasing protection for the vaquita. This move led to fishers complaining that they weren't warned of this new move [128].

The hooks catch the nets, preventing them from capturing fish or any other creatures. In Cambodia, conservationists used a similar tactic. They used bags of concrete set with rebar spikes in shallow water.

Bycatch is a leading cause of mortality for many endangered and vulnerable species, but it is more than that. Bycatch also disrupts marine food webs. It endangers ecosystems and makes them less resilient to climate change. Furthermore, bycatch is wasteful. Estimates state that fishers discard up to 40% of the global fish catch as bycatch. Millions of tonnes of marine life go to waste every year [129]. These discarded animals often die before returning to the ocean. They contribute nothing to food security. They add nothing to economic productivity. But they do add to ocean pollution. Sustainable management of these species would have significant ecological value. For example, juvenile fish caught and discarded reduce future breeding populations.

Bycatch also wastes resources used in fishing operations, including fuel and equipment. Capturing and then discarding non-target species uses time, fuel and the energy of workers. It also adds wear and tear to fishing gear. Fishers increase the number of fishing trips to meet fishing quotas. Loss and damage of equipment costs money. Lost fishing gear, such as gillnets, hooks and other gear, can continue to entrap and kill marine life for years. This is a phenomenon known as "ghost fishing." This can stop animals from eating and breeding. It can even stop animals from breathing if they need to surface in the water for air or their gills become trapped. This represents an invisible form of bycatch that is both wasteful and destructive.

Bycatch and climate change are interconnected; each amplifies the impacts of the other. Both cause loss of resilience, reduced carbon sequestration, migration of species and habitat loss. All of which affect marine environments. Juvenile animals caught and killed as bycatch stop future breeding. With no younger animals left to breed, population numbers decrease at a faster rate. These smaller populations may not be able to survive the changes caused.

Ocean acidification warms waters and reduces oxygen levels, changing habitats. Removing predators worsens the situation. It allows animals lower in the food chain to overpopulate. This then increases competition for resources, such as food. Declining resources threaten entire populations, exacerbating the problem. The depletion of resources and the ruin of habitats can cause migration. It is well known that climate change is causing species to move to more suitable environments. If the intended catch of fishers migrates, fishers will move to find them. This can lead to fishers intercepting other species that would have been safe from nets, hooks and traps.

Marine ecosystems also play a vital role in sequestering carbon, storing it in the bodies of living creatures and dissolving it in water.

Fishing techniques, such as dredging, disturb the seabed, capture unintended creatures and destroy marine plants. This further disrupts ecosystems and releases carbon as plants die and do not regrow.

Green Turtles

Green turtles (*Chelonia mydas*) are one of seven species of sea turtle. They are widely recognised for the critical role that they play in maintaining marine ecosystems. These endangered turtles face threats in many areas. Human activities, such as habitat destruction, bycatch, climate change, and pollution, have reduced their numbers. To safeguard these valuable creatures, they are protected under several international agreements. These include the Convention on International Trade in Endangered Species (CITES)[126].

Bycatch is one of the greatest threats to green turtles as they easily become entangled in fishing gear. These entrapments can lead to injuries, stress, or death by drowning when the turtles can't surface for air to breathe. The loss of individual turtles hampers efforts to restore their numbers. But the ecological consequences are also severe. Green turtles are critical to supporting the marine environment. As they graze on seagrass, they maintain the seagrass beds and help to keep coral reefs healthy. With no green turtles, seagrass ecosystems could become overgrown and unhealthy. This would have cascading effects on biodiversity, reducing the resilience of these habitats to further pressures.

Penguins

Penguins are significantly affected by bycatch. Fourteen of the eighteen penguin species get caught in bycatch events. Gillnets, trawls, and longlines are the greatest threat to these birds. For instance, Magellanic penguins in South America are frequently tangled in trawl fisheries. This causes hundreds of deaths every year, particularly in areas where they forage close to fishing operations. These deaths disrupt local populations. Having fewer birds to reproduce is reducing the number of future generations [126].

Yellow-eyed penguins, already critically endangered, are particularly vulnerable to the danger of bycatch. Set nets, used in commercial and recreational fishing, are nearly invisible underwater. As they cannot see them, penguins become entangled and drown. Bycatch has caused their population to decline by 76% over the past 20 years in some regions. There are fewer than 250 breeding pairs left in New Zealand's South Island. The situation is so dire that the loss of a few birds could have a catastrophic effect on population stability.

Bycatch reduces a population's resilience to other threats such as habitat loss, disease, and climate change. The cumulative impact can shift ecosystem dynamics. Animals, such as penguins, are a vital component of marine food webs. They control prey populations while being prey for predators. This creates a balance that is central to healthy ecosystems.

North Atlantic Right Whales

North Atlantic right whales are among the most endangered large whale species. At the time of writing, there are 366 individuals left in the wild. As with the green turtle and yellow-eyed penguin, bycatch is a significant threat to their survival. Entanglement in fishing gear often leads to severe injuries. The animals can suffer deep cuts in their flesh, starvation, or drowning when they cannot surface for air. Fishing gear and ghost nets can wrap tightly around their mouths or flippers. This restricts their movement and stops them from feeding. Fishing gear causes injuries that lead to death and serious harm. This worsens the situation for their already fragile population numbers [126].

Bycatch of right whales also has broader ecological and conservation implications. North Atlantic right whales have a very low reproduction rate. Females give birth to one calf every three to five years after a gestation period of about 12 months. The loss of even a few individuals significantly impedes population recovery. Climate change compounds this issue by shifting the distribution of their prey. Whales will move closer to areas of human activity, increasing their risk of bycatch.

These animals are also endangered by ships. These large boats can collide with whales and cause the separation of mothers and calves. Then there are whalers, those who hunt and kill whales for profit. Shipping traffic also creates a lot of noise underwater that is chronic for whales. The constant high-level sounds create "acoustic smog," making it hard for whales to communicate, navigate, or hunt [130].

A Destructive Fishing Practice

Blast fishing, or dynamite fishing, is illegal in many countries. It is a damaging fishing practice that uses explosives to kill or stun fish for easy collection. The shockwave from the explosion kills fish within a certain radius of the blast. This causes them to float to the surface, where fishers harvest them. This method is highly destructive. The explosions kill the intended fish but also obliterate the surrounding habitats [131].

Blast fishing is quick and efficient. Fishers use explosives when traditional fishing methods are less effective, like in dispersed fish populations. It is a method that provides fishers with a bigger catch with less effort. But this only works in the short term. It is often driven by poverty, lack of regulation, or the high demand for fish. But the long-term damage that it causes means that it is illegal in most countries. The explosions kill the intended fish but also destroy coral reefs. Coral reefs can take decades or even centuries to recover, but the damage can be irreversible.

Setting off explosions in bodies of water indiscriminately kills marine organisms. It disrupts and can even obliterate the balance of marine ecosystems. Species essential in maintaining healthy oceans, rivers, and lakes are often lost. This practice is also self-defeating. Repeated use of explosives depletes fish stocks, leaving nothing to catch. This harms all creatures, young and old, within the blast radius. Young animals do not live to breed, preventing future generations, so there is no ability to recover from the damage done to population numbers. The use of explosives also introduces toxins into the water, further harming marine life and degrading water quality. The long-term damage caused by blast fishing can be devastating. The explosions leave behind barren landscapes that cannot support marine life. The loss of coral reefs has severe implications for biodiversity. These ecosystems are home to a quarter of all marine species; they cannot be replaced by other habitats.

River Dolphin Conservation

In December 2021, Emily Hermann wrote about a project designed to protect river dolphins in the Mahakam River in India. For more than 25 years, net entanglement has killed two-thirds of dolphins. This continued until only 80 individuals remained. This was also a problem in other rivers in the region, such as the Ganges, Indus, and Mekong, which also had declining dolphin numbers [132].

The project trialled putting electronic devices on nets that made pinging noises. These pings alerted dolphins to the nets before they were close to danger. For six months, they were 100% effective. No dolphins were tangled in fishing nets. This was a success; not only were dolphins safe from nets, but nets were safe from dolphins.

Another benefit was an increase in fish numbers as dolphins weren't eating them, and fishers increased their daily catch by over 40%.

This was a major victory for the dolphins and the freshwater ecosystems. These animals stabilise the habitat, maintain fish populations and the river ecosystem as a whole. High numbers of river dolphins are a sign that the river is thriving and fish numbers are high. It shows that the food that fish eat, plants, plankton and smaller fish, is stable. This project aimed to minimise the damage from human activities. But it is disappointing to find a solution that allows those activities to continue.

The Pollution Left Behind

Fishing creates a lot of pollution. Plastics, oil, discarded bycatch, air pollution from the use of fossil fuels, and more. Also, fishing methods such as bottom trawling stir up sediment, resuspending it and dirtying the water.

Lost or abandoned fishing equipment, often referred to as **ghost gear**, is a significant source of marine pollution. Hannah Ritchie wrote about some of these problems in 2023, with a focus on plastics. Her article "Most plastic in the Great Pacific Garbage Patch comes from the fishing industry" (2023) gave insight into the issue. Published online at Our World in Data, she wrote, *"by mass, around 46% of the plastics were fishing nets and ropes. This was around 60% for plastics larger than 5 centimetres."* This article focused on the Great Pacific Garbage Patch, an area of concentrated waste in the central North Pacific Ocean. Created by ocean currents channelling floating waste, it settled a few hundred miles north of Hawaii. Studies on the waste found that "between **75% and 86%** of floating plastics larger than 5 centimetres could be from abandoned, lost or discarded fishing gear." [133]

Larger plastic items, especially those that wrap around objects, damage ecosystems and trap marine life. Trapped animals can suffer injuries or death as a result. Ghost gear accounts for an estimated **10% of all** marine litter by volume. It is particularly dangerous as it continues to "fish" indefinitely, trapping and killing sea creatures. The fishing industry uses equipment made from durable plastics. Items including nets, traps and lines are made of materials like nylon and can drift in oceans for decades. This allows them to continue entangling marine life. Animals like turtles, seals and birds become trapped, suffering and dying. Ghost gear is particularly problematic because it can degrade into microplastics. These tiny fragments enter food webs and persist for generations. They also affect the health of marine creatures, accumulating in the food chain. This means that people who eat fish and seafood are also eating plastic.

Nets, lines, and traps made of plastic materials persist in the ocean for decades. They will either float in the water or sink to the ocean floor. Fish can ingest small pieces of plastic, filling their stomachs. Chemicals released from these plastics can poison them. They can also break down into the smallest microplastics that enter their bloodstreams. Studies of marine species have found microplastics stuck to the gills of squids. Plastics also passed into the digestive tracts and collected in the ink sacs of the squid L. vulgaris. Microplastics have been found in the livers of European anchovies, *Engraulis encrasicolus* (L. France Collard et al). People believed that plastic eaten by fish passed through their bodies. But it is now known that microplastics can accumulate, collecting in various organs of the bodies of marine creatures. The threat to animals living in the plastic-polluted waters passes on to humans who eat fish and seafood. It is now well understood that most people on the planet have plastic in their bodies (Gong et al. 2021). Examinations have found plastic in human faeces (Schwabl et al., 2019). In their blood (Leslie et al., 2022), in the placenta (Ragusa et al., 2021), and in breast milk (Ragusa et al., 2022). There is also a link between eating plastic-contaminated fish and seafood and reduced male fertility (Gore et al.,

2015). This link proves that plastic pollution from fishing vessels poses a threat to everyone who eats the creatures they catch [134].

Fish and other marine organisms, unintentionally caught and discarded, add to marine pollution. Added to this are fish body parts. Removed during the processing of a catch on board fishing vessels and thrown overboard to rot. As these discards decompose, they can alter the nutrient balance in local waters. This can overload waters with nutrients, causing reduced oxygen, algal blooms, and eutrophication. Bycatch is a major pollutant in fishing operations. But it is unavoidable with current fishing methods.

Accidents and routine operations add to the pollution that fishing vessels cause. Spills of fuel, oil and chemicals from boats harm water quality and aquatic organisms. Smaller, consistent leaks from fishing vessels are often overlooked, but collectively, they have a considerable environmental impact. These spills contaminate water, harm fish and aquatic organisms. They lead to long-term damage to habitats like coral reefs and mangroves. Chemical releases into water, like a constant fuel leak, contribute to ocean pollution. The problem is that there are no easy ways to repair the damage.

The leaks cause an oil slick on the ocean's surface that sunlight cannot penetrate. Photosynthetic organisms like plankton, critical to the marine food web, suffer. Chemicals, like cleaning agents, antifouling paints, and hydraulic fluids, can harm aquatic life on contact. They also add toxic substances that degrade water quality and poison habitats.

Over time, persistent pollutants from fishing vessels accumulate in marine ecosystems. This leads to more widespread and chronic damage. Chemicals like heavy metals and hydrocarbons can create the same problems as plastic. They can bioaccumulate in aquatic organisms, moving up the food chain. Eating these animals poses a risk to predators, including humans. The ongoing pollution is cumulative, contributing to the broader decline in ocean health. It exacerbates issues like acidification and hypoxia in already stressed marine environments. Fishing vessels pollute the air with the burning of fossil fuels. Emissions of carbon dioxide, nitrogen oxides, and other greenhouse gases worsen climate change.

Fishing, particularly bottom trawling or dredging, disturbs the seabed, resuspending sediments. Bottom trawling, dragging heavy nets along the ocean floor, stirs up large amounts of sediment. The disturbance displaces fine particles, organic matter, and nutrients that had been on the seabed. The use of anchors and the movement of fishing vessels in shallow waters can also churn up sediments. All of this sediment disruption is damaging to natural processes on the sea and ocean floors, disturbing both plants and animals.

Resuspending sediment clouds the water. Photosynthetic organisms, like seagrasses and algae, receive less sunlight, reducing their growth. Sediments can clog the gills of fish and invertebrates, making it difficult for these animals to breathe, increasing their mortality rates. The effect is like humans trying to breathe smog-filled air. Additionally, sediments can carry pollutants such as heavy metals, pesticides, and organic waste. Stirring this can spread pollution through the water and harm marine life. Resettling sediments can smother coral reefs, seagrass beds, and other vital habitats. The effect is like the settling of ash after a fire. This further disrupts ecosystems and reduces biodiversity. The cumulative effects of sedimentation degrade water quality and habitats. Over time, it becomes harder for marine ecosystems to recover.

The resuspension of sediment reduces water clarity, impairing the ability of photosynthetic organisms, like seagrasses and algae, to access sunlight, leading to reduced primary production. Sediments can clog the gills of fish and invertebrates, making it difficult for them to breathe and increasing mortality rates, having a similar effect to humans trying to breathe smog-filled air. Additionally, sediments may carry pollutants such as heavy metals, pesticides and organic waste, which can spread through the water and harm marine life. Resettling sediments can smother coral reefs, seagrass beds and other vital habitats, disrupting ecosystems and reducing biodiversity. Over time, the cumulative effects of sedimentation can degrade water quality and habitats, making it harder for marine ecosystems to recover.

Crossing Oceans Twice

One element of the fishing industry, not a part of public discourse, is processing fish in different countries. The process is that fish, caught in one country, are shipped to another and processed before returning for sale. The UK is a prime example of this practice.

The sending of UK-caught fish overseas for processing is a striking example of inefficiency in the global food system. Fish such as cod and haddock are often exported to countries like China for filleting, canning, or freezing. The frozen fish is then shipped back to the UK for distribution and sale. Driven by low labour and processing costs, it is cheaper to process fish in this way. The practice continues despite the environmental and ethical concerns that it raises [135].

One of the major issues with this system is the significant carbon footprint it creates. Storing fish cold and transporting them thousands of miles, twice, uses a lot of energy and resources. This practice not only adds to greenhouse gas emissions. It also contradicts the principles of local and sustainable food systems, highlighting how lowering costs often takes priority over any environmental considerations.

From a sustainability perspective, this approach undermines the concept of local food. Fish caught in British waters should support local economies and minimise transportation emissions. This raises important questions about the hidden costs of cheap food and the trade-offs between economic efficiency and environmental responsibility.

For consumers, this practice often goes unknown as the packaging of fish products rarely tells the full story. While a label might state "caught in the UK", it may not state that the fish were processed thousands of miles away. This lack of transparency means consumers cannot make informed choices, stopping them from knowing about the environmental impact of the seafood they buy.

Aquaculture

Aquaculture, or fish farming, is often promoted as a solution to overfishing. Many believe that it can stop the damage done by fishing vessels. Currently, most of the fish and seafood people buy come from aquaculture. Roughly 51% of the fish people eat are farmed. In fishing, of all fish caught, approximately 60% are sold for human consumption. The remaining 40% is used for other purposes, including providing feed for fish farms [136].

Aquaculture farms a variety of aquatic organisms for commercial purposes. Common species cultivated in aquaculture include salmon, tilapia, catfish, and carp. The shellfish produced include shrimp, prawns, oysters, mussels, and clams. But crustaceans, crabs, and lobsters are also a part of this industry. Some aquatic plants are also grown, with seaweed and algae farmed for food and biofuels [137].

Being "aqua"-culture, this farming method must take place in water. But it can occur in different settings, depending on the requirements of what is being farmed. There are also limitations, such as space and access to the sea or fresh water. Indoor methods use Recirculating Systems (RAS). These recycle water through a series of water-containing tanks. These systems are a more efficient closed-loop system. They maintain water conditions while conserving resources. By filtering and recirculating water through connected tanks they minimise waste and water use [138].

One of the oldest and simplest methods of fish farming is the pond system. Ponds use freshwater for species such as tilapia, catfish, carp, and shrimp. They involve cultivating fish in controlled ponds. The ponds can be natural ponds, identified as fit for purpose, or man-made. Man-made ponds are dug out to the required dimensions and lined with earth, clay, or concrete. They are then filled with water from natural sources, such as rivers, wells, or rainwater. At harvest, fishermen catch fish with nets, or they drain ponds of water, revealing the fish. In systems that drain the ponds, water is released into the surrounding environment.

In open saltwater aquaculture, such as large lakes and seas, fish are contained in cages or net pens. This method combines the natural aquatic environment with controlled farming practices for species like salmon, tuna, sea bass, and trout.

Net pens are enclosures that are anchored to the seabed, with an open top and netted sides for water flow. Cages use the same principle of allowing water flow, but are rigid and submerged in water. The flow of water maintains water quality, with currents bringing fresh water while removing foul water and waste. The fish are harvested using nets or pump systems that collect fish into containers.

An emerging method of farming fish and seafood is Integrated Multi-Trophic Aquaculture (IMTA). This more sustainable approach to aquaculture mimics natural ecosystems. The system raises several species from different trophic levels (levels of the food chain) together. Each species plays a specific role, creating a balanced and efficient system where the waste of one species feeds another [139].

Species are chosen for IMTA based on their interactions and how one would benefit the other. Each species uses the waste of another as a food source, stopping waste from accumulating in the water. Salmon and tilapia can grow in the same environment as mussels. The mussels filter the water and remove fish waste. Nutrients excreted by the mussels and dissolved in the water feed seaweed or algae. Each of these organisms is later harvested. Of all fish farming methods, IMTA is the most complex and expensive. It requires expert knowledge to design, build, and manage. This means that the adoption of this method is slow, with few companies wanting to, or able to, invest in the systems.

Aquaculture, An Improvement on Fishing?

Aquaculture is seen as being better for fish and the environment. But aquaculture could also be better for fishers. Fishing is one of the world's most dangerous professions. A study by the FISH Safety Foundation found that there are more than 100,000 fishing-related deaths each year. From fatal injuries to decompression sickness from deep-sea dives, fishing is high-risk. Aquaculture, being more controlled, removes many of the risks [140].

Aquaculture also reduces pressure on wild fish populations, as breeding fish can prevent the declines recorded over the last 100 years. This supply is also stable as people control the number of fish farmed, increasing and decreasing numbers as required. But, while aquaculture has advantages, it also presents challenges. These include pollution, fish escape, disease outbreaks, and ethical concerns for the animals.

Farmed fish often escape from aquaculture facilities, negatively impacting wild fish populations. Escaped fish can outcompete native species for food and habitat. They can spread diseases. If they interbreed with wild fish, they can cause a loss of genetic diversity and resilience. Escaped non-native species can also cause a complete replacement of native species. An example of the negative effects of the escape of farmed fish is an incident that happened in Iceland.

Escaped Salmon

In September 2023, The Guardian released an article about the escape of approximately 3,500 salmon. The fish broke out of an Arctic Fish farming facility in Patreksfjörður, located in Iceland's Westfjords. The escape was due to two holes in the farm's net that were large enough for fish to find and swim through. These escaped salmon entered several rivers in northwest Iceland, including major salmon-fishing rivers. Escaped fish were easy to identify and distinguish from native salmon. Conditions on the farm had caused them to have "worn gill covers, shortened and disfigured snouts and missing or torn fins." Locals, finding fish in more than 32 rivers, posted photos on social media. One post had a photo of a fish covered in sea lice, a parasite that can be lethal to wild fish [141].

"This is an environmental catastrophe. If they breed, the salmon will lose their ability to survive." - Guðmundur Hauker Jakobsson

The main environmental concern of escaped fish is the genetic mixing between farmed and wild salmon. The genes from selectively bred fish could weaken the genetic integrity of native salmon populations. This mixing might reduce the fitness and survival rates of wild salmon. Farmed fish, which have been bred with wild species, produce young that mature faster. This affects the ability of fish to breed, threatening populations of wild species. Additionally, farmed salmon compete with wild salmon for food and habitats. They can introduce diseases or parasites, such as sea lice, into the wild population. This event intensified debates in Iceland about the sustainability of open-net salmon farming, emphasising the risk such escapes pose to Iceland's prized wild salmon populations. Local fishers demanded stricter regulations to prevent future escapes.

Disease

As in poultry farming, diseases are a significant issue in fish farming. This is primarily due to high stocking densities. But poor water quality and pathogens, from feed or untreated water, also contribute.

Common diseases include bacterial infections like Vibriosis. Parasites such as sea lice feed on the skin and blood of fish. Viral diseases like Infectious Salmon Anaemia (ISA) can have a 100% mortality rate. These illnesses can also cause reduced growth and financial losses. The over-reliance on antibiotics and chemicals to control diseases worsens antimicrobial resistance and environmental damage. Open-net systems heighten the risk as diseases can spread to wild fish. This impacts local ecosystems in the same way that bird flu on poultry farms infects wild birds. Preventative measures, including vaccinations, antibiotics, and improved biosecurity, help. But disease management remains a complex challenge [142][143][144].

Diseases in fish farming damage ecosystems by disrupting their balance. Wild fish have no immunity against the diseases that infect farmed fish. When they become ill, their ability to recover is poor. Additionally, the excessive use of antibiotics and pesticides has several negative impacts. Water can become polluted with chemicals, and antibiotics in the water lead to the development of antimicrobial resistance in aquatic microbes. This trifecta of disease, chemical pollution, and antibiotic-resistant microbes further harms ecosystems, putting wild populations at risk.

Pollution and Land Use

Disease outbreaks can cause an accumulation of organic waste. Dead fish and uneaten feed left in the water create the same pattern of eutrophication discussed in Chapter 3. Water depleted of oxygen causes dead zones where marine life cannot survive. These imbalances can trigger harmful algal blooms, further degrading water quality. In open net fish farming, this threatens both wild and farmed aquatic organisms. Over time, these cascading effects reduce the stability of ecosystems, compromising the sustainability of both aquaculture and traditional fisheries [137].

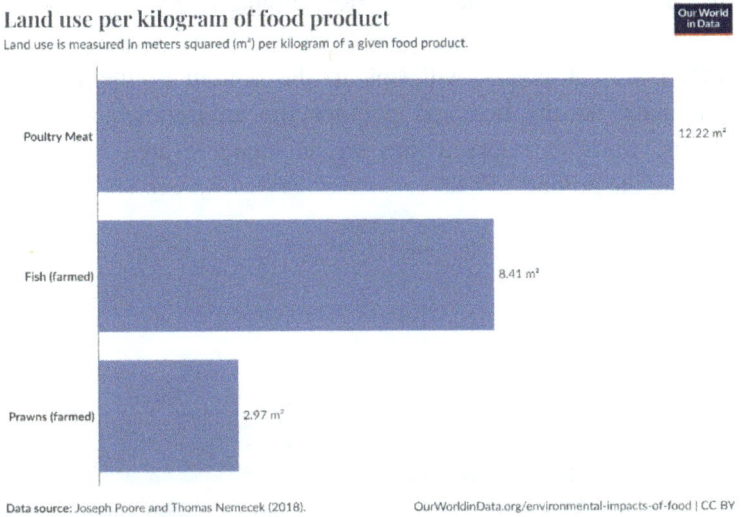

Figure 8

Fish farms, particularly in coastal areas, often destroy critical habitats such as mangroves and wetlands. Figure 8 shows the land needed for farming aquatic animals versus poultry farming. Farming fish and prawns uses a considerable amount of land, even though they are aquatic creatures. Shrimp, prawns, and milkfish are cultivated in tropical and subtropical regions. This can mean clearing mangrove forests and wetlands, causing long-term ecosystem disruption. Mangroves are critical ecosystems. They protect coastlines from erosion, provide nursery habitats for fish, and sequester carbon. The loss of these habitats reduces biodiversity as animals are killed or migrate away. The ecosystems become insecure, increasing the vulnerability of coastal communities to climate change.

Fish farming generates a large amount of waste. Uneaten feed, fish excrement, and chemicals used to treat diseases all need to be dealt with. The Norwegian Pollution Control Authority has analysed this. They found that a medium-sized fish farm of about 3,000 tonnes can produce as much effluent as a city of 50,000 people. This waste, without proper management, can contaminate surrounding waters.

This, again, brings us back to nutrient pollution, algal blooms, and degraded water quality. The waste from fish farms, whether in open water, lakes, or ponds, will sink to the bottom of the water. It accumulates at the bed and pollutes the environment. In pond systems, waste is released into the surrounding area at harvest, polluting the ground [145].

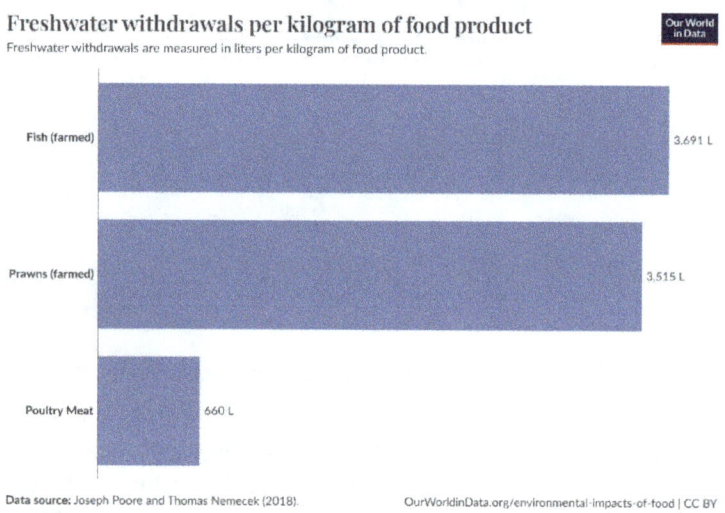

Figure 9

As in the farming of land animals, ponds used in aquaculture need to be cleaned between uses before being refilled. Figure 9 shows freshwater use in aquaculture compared to poultry farming. Wastewater will contain bacteria, fish excreta, dead fish, and uneaten rotting food. Hydrogen peroxide and pharmaceuticals given to fish, including hormones, will contaminate the water. This same mixture will remain in the mud at the bottom of the pond, along with algae, moss, and fungi. In preparation for the next school of fish, farmers cover the remaining mud with quicklime. This breaks down algae, kills bacteria, and restores minerals in the soil. The lime also increases the pH of the ground, which can be very acidic after being covered in organic waste for long periods, while fish grow. In open waters, seas, and oceans, waste washes into the surrounding environment. Antibiotics and other chemicals used in fish farming can contribute to the development of antibiotic-resistant bacteria. These bacteria can then infect wild fish, becoming a risk to human health due to their persistence.

Salmon Farming in Chile

Nearly half of all farmed salmon imported to the United States are from Chile, the world's second-largest producer of farmed salmon. In 2022, the market value of Chilean salmon exported to the U.S. was $3 billion. Most of this salmon is not considered good quality. Over two-thirds of Chilean farmed salmon are rated red by Seafood Watch. This is because of the overuse of antibiotics. Chilean salmon farms use large amounts of antibiotics to control the bacteria *Piscirickettsia salmonis*. This is an endemic bacterial disease that has a high mortality rate. To manage this, Chile uses more antibiotics than any other salmon-farming country [146][147].

> *"Salmon farming, at least done in the way it is done in Chile now, has a huge environmental impact,"* - Alan Friedlander, *a researcher at the University of Hawaii researcher and science director of the National Geographic's Pristine Seas programme.* [148]

The quick growth of salmon farming in southern Chile has led to both ecological and economic issues. High-density fish farms produce a lot of waste and leftover feed, leading to nutrient overload and algal blooms in nearby waters. Escaped farmed salmon threaten local biodiversity, compete with native species, and spread diseases. The communities of Chile are reliant on traditional fisheries, but salmon farms create economic hardships due to reduced fish stocks and environmental degradation.

These salmon farms must be constructed; they are not natural, and farmers will clear coastlines to make way for them. Entire habitats are damaged or destroyed. Marine mammals are displaced, including different species of whales and dolphins. The farms threaten many of the endangered species of Chile, including humpback whales, orcas, sea otters, cold-water corals, albatrosses and penguins. It is also documented that some salmon farm workers will kill sea lions that eat food from the salmon pens. Moreover, it is important to remember that salmon are not native to Chile. When these fish escape farms, as happens all too often, they are an invasive species. Becoming predators, they eat native species and consume the food of other creatures, displacing those that would have been in their place within the food web. They threaten both predators and prey while introducing diseases that native species have no natural protection against.

According to the Chilean campaign group, Terram, the density of salmon in cages is very high. A typical farm, with 660,000 salmon, releases roughly 879 tonnes of waste per cycle. This waste includes uneaten feed, dead fish, fish faeces, and nitrogenous waste. Aquaculture researcher Doris Soto reviewed waste production in Chilean salmon farming. She found that farms produced an average of 4,500 tonnes of salmon every growing season. They also deposited 225 tonnes of nitrogen on the seabed. The resulting eutrophication, algal blooms, and vast amounts of rotting waste have ruined the water. Nearby shellfish beds are fouled, and wild fish populations have declined.

Fish Welfare

Fish are rarely a part of the conversation when talking about animal welfare. That animals are sentient, conscious and have feelings, with wants, likes and dislikes. That fish can feel stress or suffer. In aquaculture, welfare is a growing concern, as fish are treated as commodities. The EU has some regulations addressing fish welfare, and the UK has the Animal Welfare Act. This requires that farmers meet the "Five Freedoms". Animals must be free from hunger, discomfort, pain and fear and be free to express normal behaviour. Yet, detailed considerations for fish are often lacking. This is especially true when compared to terrestrial animals (which is also poor, compared to those given to animals thought of as pets). Key welfare challenges include overcrowding, poor water quality and inhumane slaughter methods. The overcrowding on fish farms leads to competition for resources. Fish suffer physical injuries and social stress. Slaughter methods such as asphyxiation, live gutting, or ice chilling are inhumane. Also, these fish are wild animals; they are not domesticated. They do not want or understand interactions with humans. Handling during grading, vaccination, and transport is distressing for these sentient creatures [149].

Stress in aquaculture has origins beyond overcrowding and human contact. Fish naturally avoid low oxygen levels, high ammonia concentrations, and temperature changes. They are very sensitive to being removed from water. Fish drown in air as we drown in water, meaning that routine handling is extremely traumatic for them. In fish, stress can manifest in physiological changes that are not obvious. Elevated cortisol levels and suppressed immune responses harm the animals. More noticeable behavioural signs can be increased aggression, erratic swimming, or listlessness. As in humans, chronic stress weakens fish and suppresses their immune system. This makes fish more vulnerable to diseases, which are rampant in aquaculture systems.

Diseases in fish farms are common and a major welfare concern. Common diseases include sea lice infestations, bacterial infections, and viral or fungal diseases. These conditions cause lesions, organ damage, and systemic illness, and sick fish suffer significant pain and distress. Fish suffering from these diseases will isolate themselves, swim abnormally, or not feed. Farmers rely heavily on antibiotics and chemicals to control infections and diseases. But this not only compromises the welfare of fish; it harms surrounding ecosystems.

Seafood Watch is a sustainable seafood advisory list that certifies fish quality. They listed farmed salmon from several countries as red, as antibiotic residues exceed maximum levels. In the UK, only 14% of the farm-raised salmon is certified; 86% is red. In Norway, 34% is certified; 66% is red. In Canada, 3% of the farmed salmon is yellow; 94% is red; only 3% pass.

Behaviourally, farmed fish differ significantly from their wild counterparts. Wild fish travel vast distances, follow natural migration patterns and live in diverse habitats. In contrast, farmed fish remain confined to small, crowded enclosures. Their movement is severely restricted, preventing usual swimming behaviours. Socially, wild fish often form complex hierarchies. They exhibit schooling behaviour in open water. But overcrowding in farms disrupts these dynamics, leading to increased aggression and stress. Feeding behaviour also differs. Wild fish forage for food as part of their natural routine, but farmed fish are fed to a schedule. This often results in feeding frenzies and competition. As fish learn the cycle of food provision, understanding that there will be no more food for some time, they fight to eat. Reproductive behaviours are similarly affected, with hormone treatments and artificial selection manipulating the sex of fish and breeding. This strips fish of their natural mating rituals and habits. Also, in aquaculture, all the fish raised together are the same age.

In nature, fish and other creatures live in communities of different ages. Fish populations consist of individuals of various ages and sizes. They form complex social structures; older or larger fish often take on dominant roles. Younger or smaller individuals occupy different positions within the group. These hierarchies help mediate competition and maintain social order. In aquaculture, this is lost. The uniformity of age and size removes these natural structures, leading to increased competition and aggression. All fish vie for the same resources, with no older fish controlling them.

Aggression becomes a notable problem in age-uniform groups. With no clear hierarchy to mediate interactions, fish may exhibit increasingly aggressive behaviours. Chasing, biting, or fin-nipping causes physical injuries and increases stress. Raising fish in this way also stops older individuals from teaching essential behaviours. Juvenile fish do not learn foraging, predator avoidance, or social cues.

Another natural behaviour of fish that is lost in aquaculture is stress avoidance. Smaller fish often retreat to avoid conflict with larger or dominant individuals. This natural coping strategy is unavailable in uniform groups, as all fish have similar needs and behaviours. This inability to escape competition or aggression adds to the stress and frustration within the school. Grouping fish by age and size is practical for aquaculture operations. But it creates a social environment that is highly unnatural and often detrimental to the fish.

The absence of natural hierarchies causes increased aggression, crowded conditions, a lack of learning opportunities, and restricted behaviour. This all comes together to compromise fish well-being and create chronic stress. Scientific evidence has confirmed that fish are sentient. They can feel pain and experience stress, making their welfare a critical ethical concern. Despite this, fish are often overlooked in welfare discussions. Highlighting these aspects of aquaculture draws attention to the suffering of this highly exploited group of animals.

Fishing for Farming

Fishing for farming is a part of aquaculture. This makes the idea of aquaculture being the solution to fishing and the damage that it causes contradictory. Farmed fish need food; even fish kept in nets and cages in the ocean need to be fed. This food is usually in the form of a pellet. The ratio of food required for farmed animals is always inefficient. The input: output ratio is heavily weighted on the input side. In the case of aquaculture, this is called the Fish In: Fish Out (FIFO) ratio and averages 1:4.55. This means that producers need 1 kg of wild fish to produce 4.55 kg of farmed fish [150].

Fish feed requires a variety of ingredients. Chicken feathers, offal, animal blood, and chemical additives, such as pigments (particularly in the case of salmon), are all part of the mix. These ingredients are all used to produce feed pellets, but the main ingredient is fish. Each year, fishers catch roughly 90 million fish, with 20 million of them used as food for farmed fish. Using wild-caught fish to produce feed for farmed fish raises significant sustainability concerns. But this fish is also used to produce feed for other animals, including pigs, mink, and chickens.

Animal feed, made from wild-caught fish, comes in pellet form. The process starts with dehydration, removing 80% of the moisture content. The dried fish is then cooked to create a protein source. The "mash" that results is then pressed into a cake and broken down into small pieces. This is then dried again to produce fishmeal. This fishmeal, combined with other ingredients, makes fish pellets. Given this method of feeding fish, it seems ironic that many people eat fish to avoid processed food.

The reliance on forage fish for feed contributes to the overfishing of these animals. Many of these species are already heavily exploited. This is having cascading effects on marine ecosystems. The small fish used to produce animal feed are a critical food source for larger fish, aquatic mammals, and seabirds. Also, going back to the inefficiency of animal agriculture, it takes vast amounts of wild fish to produce farmed fish. For example, farming carnivorous species like salmon can need up to 1.2-1.5 kilograms of wild fish to produce 1 kilogram of farmed fish. The net protein loss highlights the lack of efficiency in using wild fish to produce feed.

This process means the environmental damage caused by fishers is part of the process of fish farming. Bycatch, burning fossil fuels, equipment production, plastic use, and the negative impact of fishing methods. They are a part of the aquaculture method. Aquaculture has been considered the future of providing fish for food. But the practice of feeding farmed fish with wild-caught fish is unsustainable.

Fish Oil

The global fish oil trade is a significant part of the fisheries and aquaculture industries. It is used in aquafeed, nutraceuticals (such as cod liver oil supplements), and pharmaceuticals. The market depends on species like anchovies, mackerel, and herring. These are primarily sourced from Peru and Chile, where fish stocks are dwindling [151]. Fish oil is rich in omega-3 fatty acids, which are essential in aquaculture for fish health. But these oils are also in high demand for human dietary supplements.

Most of the fish oil, 70-75%, feeds aquaculture. It is necessary in feeds for carnivorous species like salmon and trout. Meanwhile, human supplements use 20-25% of fish oil in products, including supplements and fortified foods. This proportion has grown over time as awareness of the health benefits of omega-3s has increased. The rising demand for human-grade fish oil has implications for sustainability. Since fish oil for human use often commands higher prices, its increased consumption can limit availability for aquaculture. The problem with this is that the aquaculture market is growing. This expansion means that the demand for fish oil is rising. This market may not be sustainable, though. If omega-3 oils are not available for the production of fish feed, the animals will become deficient. This results in lower levels of omega-3 oils in the fish that people eat. Alternatives, such as algae-based omega-3 oils and plant-based feeds, are in development. These reduce reliance on wild-caught fish for oil production. Though currently, the mainstream adoption of these alternatives is still limited. Cost and scalability challenges mean that many people cannot access them.

Aquaculture for Conservation

Aquaculture is a catch-all term for the raising of marine life in controlled farming environments. It has a broader meaning than the farming of fish for food, as it can play an important role in conservation. The goal of aquaculture for conservation is to breed endangered or threatened species in controlled environments and rebuild wild populations. For example, in the U.S., conservation aquaculture has helped restore populations of freshwater mussels. These molluscs play critical roles in filtering water and maintaining river health. Programs worldwide, including in Europe and North America, use aquaculture to conserve sturgeon. Overfishing and habitat destruction endanger this fish, valued for caviar and used in winemaking. Aquaculture is successfully growing corals for transplantation to degraded reefs. Hundreds of coral reefs are in recovery due to these conservation efforts. [152].

Restoring habitats, part of conservation, helps make environments ready for successful animal release. Having a healthy habitat enables them to live and breed in the wild. Reintroducing species that help restore ecosystems is also aiding nature's recovery. An example is breeding and then releasing shellfish, including oysters and clams. These molluscs improve water quality, filtering nutrients and sediment. It is also possible, with aquaculture, to maintain living populations of species. Keeping breeding pairs in aquaculture facilities builds a "genetic reservoir." This keeps the animals safe and helps protect against extinction.

While distinct in purpose, there is an overlap between conservation and food-production aquaculture. Conservation aquaculture methods often originate from or mirror those used in food production. Controlled spawning, larval rearing, and disease management are common processes. The difference is the harm. Food-production aquaculture can harm ecosystems. We've already discussed habitat destruction, nutrient pollution, and disease spread, threatening wild species. Conservation aquaculture seeks to do the opposite, as the aim is to repair and restore.

Climate Change and the Fishing Industry

As with poultry farming, fishing and aquaculture both contribute to and are harmed by climate change. Industrial fishing operations often involve fuel-intensive practices. Large trawlers and factory ships produce large amounts of greenhouse gases. Also, the processes disrupt marine ecosystems. They cause damage through overfishing. Destructive practices, such as bottom trawling, release carbon stored in seabed sediments, exacerbating global warming. Aquaculture, promoted as a sustainable alternative, has climate implications. This is clearly laid out in Figure 10, below. Intensive fish farming can lead to the release of **methane** and **nitrous oxide** from fish waste. The bodies of dead fish and uneaten feed will accumulate and rot in the surrounding water. Furthermore, the reliance on wild-caught fish for feed in aquaculture systems perpetuates overfishing. This continues the undermining of ocean health and its capacity to act as a carbon sink.

Both aquaculture and the fishing industry are vulnerable to climate change. Rising ocean temperatures and acidification, driven by higher atmospheric CO_2 levels, are altering fish habitats. Climate change is shifting breeding and migration patterns. This makes it harder for fisheries to find and maintain stable yields. Communities reliant on fishing suffer a lack of food and financial security. For aquaculture, warmer waters encourage disease outbreaks and reduce oxygen levels. Both result in increased fish mortality. Extreme weather events, such as hurricanes and typhoons, destroy fishing and aquaculture infrastructure. The link between climate change's causes and effects forms a cycle. This cycle endangers marine ecosystems and the future of global fish production.

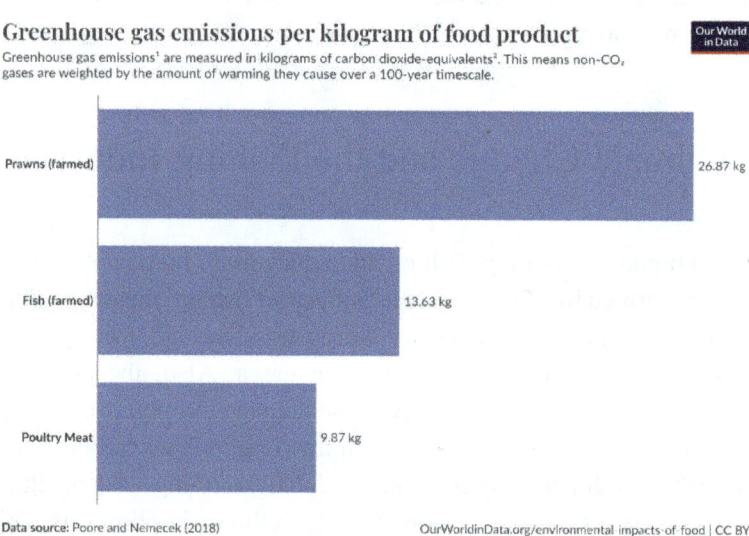

Figure 10

Migration and population decline make fishing harder. Many fishers close their businesses as it becomes too expensive to fish for small catches. Changes to farming conditions caused by climate change are challenging the aquaculture industries. Many fish species are migrating to cooler waters as ocean temperatures rise. Often travelling north towards the poles or deeper depths. This shift disrupts traditional fishing grounds, leaving some areas with small fish populations. Fishers will follow this migration as much as possible. In some circumstances, they will be unable to follow due to fish living in inaccessible areas. Fish can also cross the boundaries of countries as they move. This poses a problem: do they cross this border and potentially break the law, or do they stick to their own country? For example, cod in the North Atlantic and mackerel in European waters have migrated. Both have moved north searching for cooler habitats. These changes strain international relations, leading to disputes over fishing rights.

Some species are declining or facing extinction due to climate-related stressors. Coral reefs facing degradation, resulting from warming waters and acidification, are an example. Coral reefs serve as critical nurseries for many fish species. But they are experiencing widespread bleaching and die-offs. Fish populations that depend on these ecosystems for shelter and food are having to move or die. Atlantic salmon populations are shrinking due to warming rivers and less prey.

Aquaculture is also adapting to changing conditions. Warmer water temperatures can reduce the growth rates of certain farmed fish. Higher water temperatures increase vulnerability to diseases and foster harmful algal blooms. Some fish farms have had to move to cooler regions or deeper waters to maintain suitable conditions. As the temperature of the water goes up, fish yields go down. In warmer areas, the temperature is rising beyond the tolerance of fish. Farmers can consider cultivating heat-tolerant species, such as tilapia or barramundi, or moving. But such transitions need investment, and this may not be workable for all producers, especially small-scale farmers.

Fish farms face heightened risks of damage and even destruction in regions prone to extreme weather events. Extreme weather, like hurricanes or typhoons, causes farmers' financial losses. But they can also cause more environmental damage with broken infrastructure, released feed, and escaped or dead fish. The combination of fish migration, population declines, and worsening conditions is reshaping fishing and aquaculture. The result is often economically and environmentally unsustainable.

Tassal

Tassal Group is a prominent player in Australia's aquaculture industry. Established in 1986 to farm Atlantic salmon in Tasmania's southern waters, it has grown into a behemoth. The group is now Australia's largest vertically integrated seafood producer, managing operations from hatcheries and marine farms to processing and marketing. The group produces 40,000 tonnes of salmon per year and has also diversified into farming Australian Black Tiger Prawns. More recently, Cooke Inc., a global seafood company, acquired Tassal. This was Cooke Inc.'s first major Australian investment.

From the start, Tassal has said that it "embraces sustainability" as a core aspect of its operations. It has engaged in community and environmental initiatives and maintained certifications to meet industry standards. Yet, the model that Tassal runs is counter to the image of sustainability that it tries to cultivate.

In recent years, Tassal has faced significant challenges. Environmental stressors, including storm events and related water quality issues, have impaired operations. One notable example occurred in Tasmania's Macquarie Harbour. Between September 2023 and March 2024, reports told of mass salmon mortalities. Approximately 1,149 tonnes of salmon and trout died in the region, representing roughly 10-12% of the total stock. Poor water quality and low oxygen levels were cited as potential contributing factors.

But conditions were exacerbated by overcrowding in salmon pens, which can reduce oxygen levels further. Unable to breathe, fish stress levels increase, adding to their vulnerability to disease [153].

The specific causes of the mortality events were not fully disclosed by the companies involved, including Tassal. But the incidents highlight the strain that aquaculture operations place on delicate ecosystems. Stress from environmental conditions, combined with overstocking and the expansion of the industry in already burdened waters, has sparked debates about the sustainability and transparency of such practices. For example, Tassal faced criticism for not promptly addressing these mortality rates publicly. This underscored growing concerns about transparency, animal welfare, and environmental impacts in aquaculture.

Tasmanian residents and activists stood up against Tassal's salmon farms many times. Driven by concerns over environmental degradation and loss of community resources, they protested:

Protests in Long Bay (2022): Hundreds of residents protested Tassal's plans to refill salmon pens in Long Bay. Having been emptied, the community was happy with the closure. Environmental reports had already stressed the damage the farming caused to the bay's ecosystem, and the community objected to the restocking. The Bob Brown Foundation and other groups led protests to block Tassal's supply vessel. They showed strong opposition to Tassal, advocating for the protection of Tasmania's marine habitats [154].

Coningham Beach Rally (2023): Over 400 people gathered to protest Tassal's salmon farming in the D'Entrecasteaux Channel. Local organisations and prominent figures like author Richard Flanagan collaborated on their actions. They criticised the industry's environmental impact, citing pollution and habitat destruction. Protesters demanded that Tassal either move its operations to land-based systems or close [155].

South Arm Community Opposition (2024): Groups like Friends of the Bays and Neighbours of Fish Farming joined forces. They protested the expansion of industrial salmon farms in areas like Storm Bay. Their concerns included pollution, light and noise, and threats to endangered species like the red Handfish. [153].

Tassal has also faced criticism for its partnership with the World Wildlife Fund (WWF). The WWF allowed Tassal to display the panda logo on its products. This was meant to be a mark of environmental responsibility, but the arrangement involved payments from Tassal to WWF, totalling up to $500,000 annually. The partnership has sparked controversy, with critics arguing that it compromises WWF's credibility as an environmental advocate [156].

Environmental groups and residents raised concerns about the harm linked to Tassal's operations. They included water pollution and the creation of marine dead zones. There was also the suffocation of marine life caused by fish waste and low oxygen levels around salmon pens. These are all the result of Tassal's activities in Macquarie Harbour. They contradict the sustainability standards that WWF is expected to uphold. Many were unhappy; people wanted the agreement to end. Tasmania's peak environmental organisation requested that WWF sever financial ties with Tassal. But WWF defended its partnership, stating that it aimed to drive improvements in Tassal's ecological practices. This didn't change anyone's perception of the payments, and the association remains contentious.

Veganism for Ocean Conservation

Veganism, or not eating fish or seafood, could greatly reduce or even end the environmental and ethical issues caused by fishing and aquaculture. These industries are major contributors to environmental degradation. They overexploit natural resources and cause ecosystem collapse. A vegan lifestyle prevents these issues, removing the demand for fish and seafood while promoting plant-based alternatives. These alternatives are rich in essential nutrients. They need less land and water to produce, resulting in less pollution.

Industrial fishing practices are unsustainable. They cause overfishing, destruction of marine habitats, and disruption of aquatic ecosystems. Trawling, for instance, destroys sea beds. Bycatch kills countless non-target species, including endangered marine life and even birds. Aquaculture, often considered a solution to overfishing, presents its own challenges. These include water pollution, reliance on wild-caught fish, the spread of diseases, and invasive species. Eliminating these practices by reducing the demand for fish products would allow space for ocean ecosystems to recover. We can restore biodiversity and the balance of marine life that is so vital to the health of our planet.

Both fishing and aquaculture contribute significantly to greenhouse gas emissions. Fuel-intensive industrial fishing vessels release CO_2 from the burning of fossil fuels. Aquaculture generates methane and nitrous oxide from waste accumulation. Transitioning to a vegan diet reduces the need for fishing and fish farming, cutting the carbon footprint associated with this industry.

Often overlooked in the fishing industry is the welfare of marine animals. Frequently forgotten, they suffer stress, injury, and inhumane deaths during capture and farming. Aquaculture adds to these concerns. Overcrowding, disease outbreaks, and practices like de-beaking and de-finning add to the misery. Vegan diets spare countless aquatic animals from this suffering while promoting a compassionate relationship with nature.

Many people eat fish for omega-3 fatty acids, protein and nutrients like iodine and selenium. But these nutrients can be easily obtained from plant-based sources. Omega-3s, for example, are abundant in flaxseeds, chia seeds and walnuts. Algae-based supplements get omega-3s from the same marine plants as fish (skip the middle fish). Legumes, grains, nuts and soya products are rich in protein. We can get iodine and selenium from iodised salt, seaweed, brown rice, sunflower seeds and lentils. A well-planned vegan diet meets all nutritional needs without contributing to the environmental harm caused by fishing and aquaculture.

Fishing and aquaculture are resource-intensive. They both kill vast quantities of wild-caught fish to feed farmed species. The whole process is inefficient and unsustainable. Transitioning to a plant-based diet means a more efficient use of resources. A vegan lifestyle needs less water, land, and crops. Veganism has the potential to improve global food security. It would also protect regions where overfishing occurs to supply distant markets.

Conclusion

In summary, a shift to veganism offers a practical solution to the many issues caused by fishing and aquaculture. It promotes environmental sustainability, reduces greenhouse gas emissions, and protects marine ecosystems. It also ensures people get the necessary nutrients directly from ethical plant-based sources. This transition benefits individuals and helps create a healthier planet for future generations.

When you pay for fish and seafood, you also pay to harm marine ecosystems.

By choosing vegan options, consumers support marine conservation efforts. This includes schemes to protect areas for marine species conservation, such as national parks and conservation zones. These spaces support the habitats and migration paths of aquatic creatures. They allow fish populations to recover. They also protect other water dwellers from overfishing and bycatch.

Also, we have learned the lesson of uncontrolled fishing. Today, fisheries face restrictions, quotas, bag limits, off-seasons, and size limits, all with the aim of reducing the harms of fishing. All major fisheries are shifting away from gill nets and their high risks of bycatch, but these nets are still used in less developed countries. Despite these efforts, overfishing and destructive fishing methods persist. Economic pressures, inadequate enforcement, and global demand for seafood continue to drive the sector. Veganism solves many of the environmental challenges caused by fishing and aquaculture. It also ensures that people receive essential nutrients from ethical plant-based sources. Veganism creates a future where marine life and human health coexist in harmony.

Chapter 6: Burning Forests for Beef

An Introduction to Beef Farming

Beef farming is one of the most environmentally impactful forms of agriculture. It serves as a staple in many diets around the world and is gaining popularity in many markets. However, the production of beef comes with substantial environmental costs. These include significant greenhouse gas emissions, deforestation, extensive land use and water consumption. Additionally, the intensive farming practices used in beef production contribute to biodiversity loss, soil degradation and the pollution of air and water sources.

The Process

Controlled Breeding

The process of commercial beef farming begins with controlled breeding. This typically involves the use of sperm banks that sell frozen semen from selected bulls, often online. Farmers buy semen based on desirable traits, such as rapid growth, meat quality or disease resistance. They use **artificial insemination (AI)** to impregnate cows, allowing precise breeding with no need to keep bulls.

The collection of semen from bulls for artificial insemination is a managed process. Breeders train bulls to mount a dummy cow (an artificial stand) to allow semen collection. Once mounted, the bull is guided by hand into an **artificial vagina**. This is a specialised device that mimics the temperature, pressure and texture of a cow's vagina. A restrained live teaser cow can stand by the dummy cow, encouraging the bull to mate. But in commercial operations, teaser cows are rarely used due to the risk of injury and the stress the process causes the animals. The device collects semen in a sterile tube when the bull ejaculates into the device. This sample goes to a laboratory for evaluation. Technicians check for volume, sperm motility (movement), concentration and quality.

Following testing, the semen goes through a process of preparation and preservation. An extender solution, containing nutrients, antibiotics and a cryoprotectant, preserves sperm during freezing. It also increases the volume of the sample for multiple uses. The sample is then divided into small straws, each containing a single insemination dose. The straws are then frozen in liquid nitrogen at ⁻196°C for storage.

Female cows are usually between 15-24 months of age for their first pregnancy, depending on the breed. To perform artificial insemination, the female cow (or heifer) must be in oestrus. In this phase of her reproductive cycle, she is most fertile. Hormonal synchronisation programs are also used on some farms to control oestrus. This enables the insemination of all cows while making the process more efficient.

At the time of insemination, farmers thaw the semen straw in warm water. Technicians use a long sterile insemination rod (or AI gun) to deliver the semen into the cow's reproductive tract. To guide the rod, the technician inserts one hand, up to the elbow, into the cow's rectum, to hold the cervix through the rectal wall. The rod is then passed through the vagina and cervix to deposit the semen in the uterus. After insemination, they monitor the cow for signs of pregnancy. This is usually confirmed through ultrasound or palpation about 30 days later. Artificial insemination is widely used in cattle farming, as it is cheaper and facilitates breeding on a larger scale.

Pregnancy & Growth

Once impregnated, the gestation period is the same as that of humans, approximately nine months. Both female calves (heifers) and male calves (bulls) are born, but their roles in the beef production system differ. In most cases, farmers castrate male calves, turning them into steers. This is to improve meat quality and reduce aggression. Farmers will raise a few as bulls for breeding purposes and keep females for future breeding, or raise them for meat. Farmers select animals with the best genetics or growth potential for breeding. Those with inferior traits move to beef production.

After weaning, usually at 6-8 months old, the calves move to feedlots or pasture systems. Feedlots are facilities where cattle feed on a high-calorie diet, often made from soya. Feed will also include grains to promote rapid weight gain and marbling in the meat. They remain at the feedlot until they reach their target slaughter weight, usually around 554–635 kg. At this point, the cattle go to slaughterhouses for processing. They separate their hides, meat and other body parts for uses, such as using hooves for gelatine and hides for leather.

Slaughter

The process of slaughtering cattle is as streamlined as possible. Cattle, transported to slaughterhouses, stay in holding pens to reduce stress before slaughter. Reducing stress is an animal welfare measure. But it also helps maintain meat quality, improving the taste of the meat. At least this is what people claim. At slaughter, the animals move into a stunning area. Common practice is to use a **captive bolt gun** to render the cow unconscious. The gun delivers a strong, targeted blow to the head. In many countries, stunning is a necessary step to follow animal welfare regulations.

After stunning, workers suspend the animal upside down by its hind legs. A hoist system carries the animal for **the throat to be slit**, severing the major blood vessels in the neck. This process, called **exsanguination**, allows the blood to drain from the body. This is essential for food safety and meat quality. Draining the blood reduces bacteria growth, prevents spoilage and ensures a clean carcass.

After bleeding out, there is the removal of the **hide (skin)**. This is often processed further for leather production. The **internal organs (viscera)** are then removed, including the stomach, intestines, liver, lungs and heart. Some organs are used for human food, like liver or tripe. Others go into pet food or industrial products.

The carcass is then split in half lengthwise along the spine using large chainsaws. The halves are often washed or treated with hot water or antimicrobial solutions to reduce bacteria. The carcasses are then moved to a chilling room for cooling. This step inhibits bacterial growth and improves meat texture. Once cooled, bodies go through cutting, deboning and packaging, producing pieces of meat for sale. The entire process, from breeding to slaughter, is industrialised, maximising efficiency and profitability, while meeting consumer demand for beef.

Feedlots

Feedlots, also referred to as **concentrated animal feeding operations (CAFOs)**, confine cattle. Animals feed on a high-energy diet to speed up their growth before slaughter. These operations, designed for extreme efficiency, house thousands of animals at high density in a relatively small area. Cattle are typically fed a diet of soy with grains, such as corn, along with supplements to promote rapid weight gain and produce marbled meat. While feedlots maximise productivity, their intensive nature has significant environmental impacts, particularly greenhouse gas emissions. In the United States, high-density beef feedlots are responsible for more than 7% of all emissions each year [157].

One major source of greenhouse gases from feedlots is **methane**. 32%-87% of manmade global warming is the result of cattle farming, as they emit this potent greenhouse gas during digestion. [158]. Cattle are ruminant animals; they digest food by a process called **enteric fermentation**. In this process, microbes in the stomach break down food and release methane as a by-product. This is then expelled by the animal, mostly through belching [159].

Another significant contributor to the methane emissions of beef farming is **manure management**. In feedlots, cattle waste accumulates in confined spaces at a rapid pace. It is often collected and stored in large lagoons or pits. As the manure decomposes, it releases more methane due to conditions being **anaerobic** (oxygen-deprived). The breakdown of **nitrogen** in manure releases **nitrous oxide**, another potent greenhouse gas. Levels of nitrous oxide increase again when farmers spread manure on fields as fertiliser.

Feedlots also contribute to greenhouse gas emissions through their reliance on industrial agriculture to produce cattle feed. The cultivation of feed crops like corn and soy uses large amounts of synthetic fertilisers, which also release nitrous oxide during application. Additionally, the energy-intensive processes of planting, harvesting and transporting feed, often across countries and continents, need fossil fuels. Burning fossil fuels for power adds carbon dioxide emissions to the footprint of feedlot operations.

These facilities are the prime example of the environmental costs of intensive meat production. They concentrate emissions from the animals, their waste and the feed production system. The damage that results has far-reaching impacts on climate change.

Grass-Fed Beef

Pasture-raised, or **grass-fed beef**, refers to cattle that graze on open pasture for part of, if not all, their lives. This method is often promoted as a more natural and sustainable alternative to feedlot operations. It aligns with the animals' natural grazing behaviour, and some say it supports ecosystem health. There is also the idea that grass-fed beef is better for human health, due to the nutrient content of the meat. Grass-fed beef is generally higher in omega-3 fatty acids, CLA, fat-soluble vitamins A and E, various antioxidants and potentially some minerals. Yet, the difference is small for those eating a balanced diet. For those eating a more plant-heavy diet, such as a flexitarian, the difference may be negligible [160].

The definition of grass-fed beef varies by country and certifying bodies. In some countries, animals will graze their whole lives, in others, it is for a short time. In some cases, it may refer to cattle that are only "finished" on grass as they approach slaughter weight. Although considered more natural, pasture-raised systems are not problem-free for the environment. Grass feeding produces large amounts of greenhouse gases, though impacts can differ from those of feedlots.

All cattle emit methane during digestion through the process of enteric fermentation. Cattle grazing on pasture consume a diet of grasses and forage. These are more fibrous and take longer to break down than grains, producing more methane. Grass-fed cattle also have a slower growth rate, meaning that they live longer. This results in increased methane production before the animal reaches slaughter weight [161].

As in feedlot systems, manure management is a major source of emissions in pasture-raised beef. When cattle graze on pasture, their manure deposits onto the land. This can have some ecological benefits, such as fertilising soil. But the negative is that it results in the release of greenhouse gases. As manure decomposes, it releases methane and nitrous oxide. The amount of these gases released depends on many factors. Conditions like soil moisture, temperature and manure management practices influence how manure decays. If pasture manure is not managed (e.g., it accumulates in areas or is not integrated into the soil), emissions can increase.

> *Performance: Growth rates have been excellent to date with post weaning live weight gains in excess of 1.2 kg/day. Currently, eight-month-old calves are eating 6.7 kg concentrates/day and are on target to achieve 260 kg carcass weight at 12 months of age. – The Department of Agriculture, Environment and Rural Affairs*
> *(163)*

Pasture-raised beef systems may also cause greenhouse gas emissions through land use changes. As animals move across the ground, they trample on plants, breaking them down at a quicker rate than usual. This reduces the carbon sequestration power of soil. They also disrupt soil microbe activity, decreasing soil quality [162]. Cattle also use large areas of land when grazing. The clearing of forests or other carbon-rich ecosystems can be used to create this pasture.

Deforestation releases stored carbon dioxide into the atmosphere and reduces the land's ability to sequester carbon in the future. Additionally, overgrazing in poorly managed systems can degrade soil health. It loses the ability to store carbon, further contributing to carbon in the atmosphere.

Monopolising Land

Beef farming requires vast amounts of land, both for grazing cattle and for growing feed crops. The grazing of cattle occupies around 60% of global agricultural land, yet this system provides only a small fraction of the world's calories [164]. Farmers set aside large areas of ground for grasslands to support big herds, especially in pasture-raised systems. Even in feedlot systems, where cattle feed on soy and grains, land use is enormous. Housing animals uses land. Growing crops uses land in the beef process. This extensive land use is one of the most significant contributors to the environmental impact of beef farming.

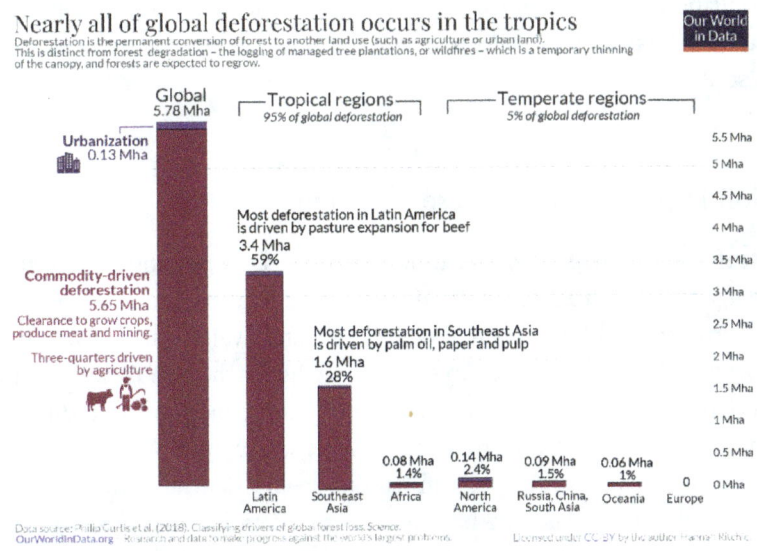

Figure 11

Deforestation for Grazing and Feed Crops

In many regions, forest clearance creates pasture to grow feed crops. In tropical regions, like the Amazon rainforest, vast areas of forest have been cleared, as shown above in Figure 11. The process of purging the land removes trees and plants. This releases massive amounts of carbon dioxide stored in trees, plants and soil into the atmosphere. These releases increase again if the disposal of trees and plants is through burning. When trees and plants are killed, they no longer sequester carbon dioxide from the air through respiration and growth. The loss of forests also destroys habitats for countless species. Animal deaths and forced migration drive biodiversity loss. Once cleared, the land is often converted to pasture. These fields remain productive for a short time before overgrazing or nutrient depletion leaves them sterile. This forces farmers to start again, to clear more forested land.

The Amazon rainforest, often called the "lungs of the Earth," is one of the regions most affected by beef farming. Large swathes of the Amazon are cleared each year for cattle ranching. This is contributing to a significant loss of biodiversity. The release of carbon dioxide from trees is impacting not only the local region but also the planet.

Soil Degradation and Desertification

Beef farming degrades soil in many ways. Overgrazing, where too many cattle graze a piece of land, strips vegetation and exposes soil to erosion. Without plant cover, wind and water carry away the topsoil, the most fertile layer. This reduces the land's ability to support future crops or pastures. Cattle trample the ground, reducing the soil's ability to absorb water, exacerbating runoff and erosion. Over time, these processes can lead to **desertification**, where once-productive lands become barren. This problem is acute in semi-arid regions, where grazing pushes ecosystems beyond their ability to recover.

Creation of Mono-crops

Farming vast fields of the same plants, primarily corn and soy, is necessary for growing crops for cattle feed. These crops are usually grown in **monocrops**, where large fields grow a single plant. Monocropping depletes the soil of the nutrients each plant needs. As the same crop is grown many times, the same nutrients are consumed, leaving an imbalance in the soil. To ensure that yields don't decrease, farmers rely on heavy use of synthetic fertilisers. Chemical pesticides are also often needed, as the lack of biodiversity makes the monocrop plants more vulnerable to pests and diseases.

The reliance on monocropping disrupts natural ecological balances. Increasing amounts of human intervention ensure continuous crop production to feed the beef feedlot system. But each intervention furthers the contribution of beef farming to environmental damage.

Consuming Water

Beef farming is one of the most water-intensive forms of agriculture. There are some who argue that the water use from beef farming has a net neutral effect. They reason that water consumed by cattle is urinated out. This argument does not work. Taking fresh water and replacing it with urine is not a net neutral; it is a net negative. The argument neglects the freshwater used on farms for cleaning and other processes. The argument also neglects the extensive water use in producing feed for cattle. Both growing and processing the crops to make feed are thirsty procedures. The urine argument disregards the exorbitant use of freshwater, making it unavailable for other uses, such as drinking. The urine argument does not consider water pollution from manure, fertiliser or pesticide runoff.

The water footprint of beef is higher than that of other animal products and much higher than that of plant-based foods. Producing one kilogram of beef requires around 15,000 litres of water. This figure accounts for all stages of production from feed crop irrigation to the processing of meat. This could be compared to the 1,800 litres needed to grow 1 kg of soya. In regions where water resources are already stressed, the high water demands of beef production can exacerbate water scarcity, impacting both local communities and ecosystems. This is particularly worrying in areas where beef production is expanding, at the expense of other land uses [60] [61].

Drinking Water

On average, a cow drinks between **30 and 50 litres** of water a day. This amount can increase to **150** litres for lactating cows or animals kept in hot conditions [163]. In **Concentrated Animal Feeding Operations (CAFOs),** large numbers of animals are confined to maximise production efficiency. Small CAFOs can house anywhere from 100 to 500 animals. A medium-sized CAFO might range from 500 to 2,000 animals. Large CAFOs, especially those used for beef cattle, can house 10,000 to 50,000 animals or more. **Very large beef CAFOs** can sometimes hold upwards of **100,000 cattle**. This is less common and is usually only found in **highly industrialised regions**.

Mega-CAFOs are among the largest facilities, housing more than 100,000 animals. Some **mega-feedlots** in the U.S. (like those in Texas, Nebraska and Kansas) can hold upwards of **200,000** cattle for beef production at any one time. If we take an average of 40 litres of water per day per cow, the smallest beef CAFO, with 100 animals, would need **4,000 litres** of fresh water for drinking per day. A large CAFO with, say, 30,000 head of cattle would need **1,200,000 litres** of fresh water every day, for drinking. This does not account for any water used to complete farm operations.

Sanitation

Sanitation is a vital part of everyday biosecurity on farms to protect both animals and humans. The processes use huge volumes of water to clean and sanitise areas.

The general order for the sanitation process is:

1. Dry clean (remove solids)
2. Wet wash
3. Rinse
4. Dry
5. Disinfect

The amount of water used for cleaning barns and equipment varies according to the size of the farming operation. Wet washes use detergents to cleanse areas housing cattle and spaces they move through. These washes are thoroughly rinsed to remove all chemical cleaners. Disinfectants, often water-based, kill microbes, fungi and viral agents that remain after the wet wash and rinse. A farm might use thousands to hundreds of thousands of litres of water per week to manage biohazards.

Slaughtering and Processing

Beyond the farm, cattle are made into meat products in slaughter and beef processing plants. The production of beef uses less water than the slaughter and processing of poultry birds. The animals are not stunned in water, and whole hides are removed manually, with no hot water required. The volume of water needed, though, remains extremely high, contributing to the water footprint of beef.

Water is used for many stages of slaughtering and processing. Following this, it is used to preserve the cold chain needed to ensure the safety of meat products. It is also used to clean, sanitise equipment and wash down carcasses at regular intervals. At slaughter, huge volumes of water wash away the blood drained from animals during exsanguination. After slaughter, carcasses are typically sprayed with hot water or antimicrobial solutions to reduce the risk of pathogens like E. coli or Salmonella. These use hundreds of litres of water per animal, depending on the facility and its practices. Additionally, wastewater treatment facilities at processing plants consume large amounts of water to manage and treat the effluent generated during these operations.

The amount of water used in the slaughter and processing stages varies depending on the size of the operation, the efficiency of the facility and local regulations. Studies have estimated that each body requires between **400 and 1,000 litres of water** during slaughtering and processing [165].

The large volume of water required for slaughter and processing can strain local water supplies. This is worse in regions already experiencing water scarcity. Additionally, the wastewater will usually contain organic matter, blood, fat and cleaning chemicals. These remains can pollute municipal water supplies and nearby water bodies if not properly treated.

Growing Feed Crops

The volume of crops grown to feed beef cattle is beyond imagining. Their production uses immense quantities of land, fertiliser, pesticides and water to grow. The amount of water needed for the irrigation of pastureland depends on the crop type, soil conditions and climate. Studies have shown that water consumption is approximately **21,000 to 34,000 litres per acre per day** [166]. It is possible to estimate the amount of water used in the feed of each animal using the moisture content of the feed and the water required to grow the crops. Cattle are typically fed a mixture of soya (10–20%) and corn (30–60%). Other ingredients, including grass, silage, alfalfa, minerals and additives, make up 20–60%.

The moisture content in various types of feed also varies. For example, silage can contain 60-70% moisture, while hay may have roughly 10-15% moisture. Grains have a lower moisture content, around 10-14%. This means that the water footprint for producing feed varies depending on the mix of the feed. For example, producing 0.45 kilograms of corn may use 378-757 litres of water, depending on irrigation methods and local climate. How this water use then contributes to the water footprint of the beef depends on how much of the corn is in the feed mix.

A mature beef cow consumes about 2-3% of its body weight in feed daily. For instance, if a cow weighs approximately 545 kilograms, it may consume around 11-16 kilograms of feed per day. This means the water used for the feed of each cow per day could be over a hundred gallons [167].

Growing crops for cattle also results in runoff. Fertilisers, organic and non-organic, become concentrated in soil, leading to leaching of nutrients into waterways. This results in the same contamination, eutrophication and dead zones that we explored in Chapter 3, which were repeated in Chapter 4 and seen again in Chapter 5.

Lagoons of Waste

Cattle, farmed in the billions each year, are large animals. This means that any harm that results is multiplied by the scale, which includes the problem of waste. On average, a single cow can generate around **24-70 kg of manure per day**, depending on its size, diet and living conditions. This makes managing the waste a major challenge in beef farming [168]. This accumulates as tens of thousands of kilograms of waste are produced per cow per year. With billions of cattle raised for beef globally, the sheer volume of manure is a huge logistical problem, and not managing this problem results in severe environmental issues.

In feedlot systems, waste accumulates in the confined areas that house cattle. It is often managed by collecting manure in lagoons or piles, concentrating it in one area. In lagoons, liquid waste is contained, while solids are composted or spread on farmland as fertiliser. When stored in piles, liquids containing pollutants can run off. Carried through the pile by gravity, it seeps into soil and waterways. Spills, runoff and contamination of nearby water sources with **nitrogen, phosphorus**, and pathogens like **E. coli** are a known problem. Pollution from manure, as with other animal wastes, can cause harmful algal blooms in rivers and lakes. This brings us, again, to eutrophication and harm to aquatic ecosystems. In pasture or grass-fed systems, manure spreads as cattle move around while grazing. Excessive amounts of manure in small areas, or overload from large herds, still lead to nutrient runoff and soil degradation.

Methane, Nitrates and Other Emissions

Methane is one of the most troubling emissions from beef farming. It arises not only from enteric fermentation in digestion, but also from manure. When stored in lagoons, anaerobic conditions catalyse the action of microbes breaking down organic material. This activity results in the release of **methane** in significant quantities. Microbes in soil produce **nitrous oxide** when manure decomposes after fertiliser application. These releases increase in wet or compacted soils. Nitrous oxide is a greenhouse gas that has roughly 300 times the potency of carbon dioxide. The greenhouse gas emissions from beef farming, particularly in such huge quantities, have a significant impact on the warming of the planet.

Beyond waste, beef farming impacts emissions through **deforestation** for grazing or feed production. When farmers clear forests, the carbon stored in trees, plants and soil is released into the atmosphere. The reduction in plant life means this carbon cannot be captured and stored within new flora. The production and transportation of cattle feed burn fossil fuels, further adding to the carbon footprint. The soya, grown for feed in tropical regions, can travel many thousands of miles to feed cattle on other continents. Every mile travelled, whether by road or sea, contributes to the carbon emissions of beef.

Ammonia is another pollutant emitted from manure. While not a greenhouse gas, ammonia contributes to air pollution through acid rain. Ammonia also damages ecosystems and human health. In sum, cow waste in beef farming presents serious environmental challenges. From nutrient pollution to significant greenhouse gas emissions. From Methane to nitrous oxide, the primary contributors to global warming. Then there are other pollutants like ammonia and carbon dioxide from land-use changes and feed production. There is nothing good that comes from treating animals and the planet like this.

Biodiversity Loss

The industry of beef farming contributes to biodiversity loss on the farm and beyond. Inside the farm, selective breeding and genetic engineering mean the loss of genetic diversity. While outside of the farm, habitat loss and damage destroy the homes of wildlife and destabilise ecosystems.

Habitat Destruction

Conversion of forests, wetlands and other natural habitats to grazing or crop fields destroys habitats critical to many species. This habitat destruction is one of the leading causes of biodiversity loss globally. The loss of habitats threatens the survival of individual species and disrupts entire ecosystems. These losses reduce their ability to function. Processes such as water purification, carbon sequestration and climate regulation slow [169].

When expanding into wild areas, beef farming often leads to conflicts with wildlife. This is a serious issue in regions where natural habitats are being converted into agricultural land. For example, in many parts of the world, large predators such as wolves and jaguars are killed to protect livestock. This leads to declines in their populations as breeding animals are the main victims. Any young they were caring for often die as a result. Farmers deal with any animals that prey on cattle or damage farm infrastructure in the same way. This can include shooting elephants that break farm fences.

The loss of predators can have cascading effects on ecosystems. These animals play crucial roles in maintaining the balance of ecosystems, controlling populations of herbivores and other species. Also, within different predator species, having a diverse range of animals at the apex of the food chain prevents the dominance of any one species.

Genetic Diversity and Livestock Breeds

The industrialisation of beef farming has led to a decline in the genetic diversity of livestock breeds. The focus for meat production is on a few fast-growing, high-yielding breeds. This has resulted in the loss of many traditional and indigenous breeds. Breeds that are often better adapted to local environments and more resilient to diseases. These new, man-made breeds are often more vulnerable to diseases and environmental changes. This further reduces the sustainability of beef farming, as animals need increased human interventions to survive. Climate change has worsened the situation by reducing suitable land for farming [170].

Another danger resulting from the loss of genetic diversity is the passing on of genes that would be considered "weak" or "inferior". Farmers work to ensure a certain amount of genetic mixing within cattle. They will make sure that a bull cannot be the father of both his children and grandchildren. Farm workers keep records of animals used and the calves produced. These checks matter in artificial insemination. If samples remain, a bull's sperm breeders can still sell and use them after he dies. The repeated breeding of a few bulls shrinks the gene pool. This is what happens when farmers select animals for desirable traits, such as high muscle gain and efficient feed conversion.

Genetic engineering of cows is an expanding area within cattle farming, with animals being designed for greater weight gain. Farmers also breed cattle for specific qualities within their meat. Newer technology allows for breeding animals to produce more male or female calves, depending on whether they want them for beef or dairy [171][172]. These genetically engineered animals are a lottery. No one knows whether the experimental animal will have the desired traits. Calves born with genetic defects will pass these on to their offspring, sometimes in place of the desired traits. Scientists have also found that genetically engineered calves are larger at birth. This often means that cows cannot have natural births; they must undergo a caesarean. This is hazardous to cows and very expensive to undertake. Genetic engineering, even with the use of technologies like CRISPR, is unpredictable. But with the desire to breed cattle that produce more meat and burp less, it is a practice that is set to continue.

Monopolising the Market

85% of the Beef Market in the U.S. is Controlled by 4 Companies

In the United States of America, four companies control almost the entire beef market - Cargill, Tyson Foods, JBS and National Beef. This is a result of decades of mergers, acquisitions and vertical integration within the meatpacking industry. These companies control roughly 85% of the beef processing market. They now have significant influence over the supply chain, pricing, production practices and lobbying power [173].

How Consolidation Happened

The consolidation of the beef industry began in the late 20th century. Larger companies acquired smaller, independent processors, expanding their market share. The use of, at the time, advanced slaughtering and processing technologies allowed these corporations to quickly scale up operations. They outcompeted smaller facilities, forcing them to sell their businesses or close. They used vertical integration to control several stages of production. Having command from feedlots to distribution, they further confirmed their power over the sector. By the early 2000s, the top four companies accounted for the vast majority of beef processing, all but removing competition in the market.

The "Big Four" beef processors' domination of the market drew attention from regulators and lawmakers. Antitrust laws, such as the **Sherman Act** and **Clayton Act**, had not been enforced, allowing the takeover. These acts exist to prevent monopolistic practices. But they were not applied to the beef industry, allowing a small number of companies to control the sector. Investigations into price-fixing and market manipulation have resulted in occasional fines. But these penalties are often too small to deter future malpractice. Cargill, Tyson Foods, JBS and National Beef use their influence to make huge profits before anyone acts to stop them.

Their actions during the COVID-19 pandemic are an example of how they operate. Ranchers were receiving historically low prices for cattle throughout the pandemic. At the same time, consumers faced skyrocketing beef prices. The Big 4 claimed that a restricted market caused both price changes. Both ranchers and customers made accusations of price gouging. In response, in 2021, the U.S. Department of Justice investigated potential anticompetitive behaviour in the meatpacking industry. The Biden administration proposed measures to increase competition that included funding for small and independent processors. But, due to their vertical integration and political influence, these companies continue to have almost complete control of the beef market.

Impacts on Farmers

Farmers and ranchers can no longer set prices for their cattle due to limited options for selling. This leaves them feeling that they have no control over how they operate, and they are often forced to accept the lower prices dictated by these corporations. This is a direct consequence of their monopolising the market. With fewer buyers in the market, producers have less bargaining power. Many smaller operations struggle to stay profitable, relying on government subsidies to make a living. The problem is ongoing, and in 2023, 14,700 American cattle farmers either closed their farms or went bankrupt. The U.S. has lost 44% of its family-operated farms since the 1980s [174].

For many decades, farmers have been feeling the influence of the Big 4 companies squeezing their margins. In 1970, farmers received 0.60 cents for every dollar of beef sold; as of 2022, this had reduced to 0.39 cents. Yet the price that the consumer pays for meat continues to increase. This price control is partly achieved by these companies using **Alternative Marketing Agreements (AMA)**. The way it works is simple. Buyers visit farmers and agree on a price for their cattle before they get to auction. Although these prices are often lower than the farmer would want, or would even get at auction, they feel pressured to accept the offer. They cannot afford to risk the possibility of being effectively blacklisted in the future. This situation has driven some ranchers out of business and further consolidated cattle production into larger industrialised systems that favour economies of scale. Those left with no livelihood feel that these companies have destroyed their lives.

Impacts on Consumers

For consumers, this consolidation is only bad. It reduces choice and creates vulnerability to price manipulation or supply disruptions. Cargill, Tyson Foods, JBS and National Beef control the price and availability. Large companies argue that their efficiencies keep prices stable. But it is obvious that reduced competition slows innovation and pushes prices up. The COVID-19 pandemic also showed how having a few companies control supply can cause low stock. Throughout this period, lockdowns and high infection rates caused supply chain disruptions. The lack of availability of meat products revealed how concentrated control can lead to supply shortages and high prices.

These corporations are also able to dominate the labelling and marketing of meat products. When a phrase, gimmick or misleading claim, such as "grass-fed" or "sustainable", becomes popular with consumers, appropriate labels will be applied to the meat product. This makes the consumer believe that a particular piece of meat is better than one without the label. These claims and phrases are not in any way regulated and have no true meaning. For example, the term "sustainable" is not defined. When applied to a pack of meat, it indicates nothing about how the meat was produced. It certainly does not mean that it is in any way different to the meat with no label, although the meat with the label may be priced 10% higher. To date, there is no "sustainable" way to produce beef, but the consumer does not know this. These marketing tricks make it harder for consumers to make informed choices, as the truth of where the beef comes from is hidden from them. This same strategy packages meat products under different brand names, even though they all have the same source.

Environmental and Ethical Implications

Consolidation also centralises environmental and ethical concerns. These corporations often prioritise maximising production over sustainability. Animal welfare is rarely considered. Large feedlots and processing plants greatly add to greenhouse gas emissions, water use, and pollution. Having a few companies dominate the market makes it hard to address systemic issues in production practices.

The greenhouse gas emissions from beef production account for 7% of the U.S.'s annual emissions and a large share of all agricultural emissions. These include methane from enteric fermentation and nitrous oxide from fertilisers. The concentration of operations in large feedlots exacerbates emissions. The production of feed for cattle drives deforestation. Companies like JBS have a responsibility here. They have been directly linked to deforestation in the Amazon, where land is cleared for cattle grazing or to grow soya for feed. This way of farming is harming consumers, farmers, animals and the planet.

Consolidation of the beef industry also intensifies ethical concerns, as profits matter most. Farmers house animals as densely as possible, with little consideration for their welfare. Industrial feedlots, or Concentrated Animal Feeding Operations (CAFOs), prioritise efficiency. Issues include overcrowding and overuse of antibiotics in disease control in unsanitary environments. Lack of competition means alternative systems, such as regenerative farming, can't compete on price.

Worms in cattle

As the winter approaches and cattle are being housed and dosed, it is important to use the correct product to prevent Type II Ostertagiasis.
Any worm larvae eaten after September do not go through normal development. These larvae enter the glands in the stomach, where they remain dormant over the winter. These are known as inhibited larvae. In incorrectly dosed animals, thousands of these inhibited larvae simultaneously develop to adults and emerge from these glands in early spring. – The Department of Agriculture, Environment and Rural Affairs (175)

The U.S. beef industry isn't the only example of market consolidation. Similar patterns exist globally, as major corporations like JBS expand their operations globally. In Brazil, Australia and parts of Europe, large meatpackers control significant portions of the market. This market practice is also applied to other meats, including fish.

Natural Nutrients

Meat, red meat, in particular, is believed to be something that humans must eat to maintain strength, protein and minerals. This idea rarely takes into consideration how the protein and minerals became a part of the animal's body. Cattle raised for beef, particularly those raised in CAFOs, feed on a manipulated diet of soya, corn and other foods. These are not a natural part of a cow's diet. Soya is used for its protein content. For the mineral content of meat, animals are often given supplements to enhance their growth, health, and productivity. These additives also ensure the nutrient content that people desire. The specific supplements vary depending on the farming system, e.g., confined farming practices vs. pasture-raised. But they generally fall into the same categories:

1. Vitamins and Minerals

- **Minerals**: Calcium, phosphorus, magnesium and potassium are often added to balance deficiencies in feed. Trace minerals like zinc, copper, selenium and cobalt are also included for immune support, reproduction and health [176].
- **Vitamins**: Vitamin A (important for vision and immunity), Vitamin D (for bone health) and Vitamin E (an antioxidant) are commonly supplemented, especially when cattle are on grain diets or during the winter when grazing is limited.

There is a common belief that cows consume vitamin B12 supplements. This is for the health of cattle and to ensure high levels for consumers and their requirements. But this is not entirely true. Cows, like humans, cannot make vitamin B12 themselves, as it is a compound produced by microbes. In the case of humans and some other animals, such as rabbits and apes, microbes produce B12 in the gut. The area of the gut where they do this is too low down in the digestive tract to allow absorption, and it is pooped out. Some animals will then eat some of their faeces to get the vitamin, but neither humans nor cows do this. People, through history, got B12 by drinking water containing B12-producing microbes, along with the vitamin. Today, with the use of sanitation, water no longer has any microbial or vitamin content, so people eat the flesh of animals or take supplements.

For cows, microbes in the rumen produce the vital vitamin with the use of cobalt. Without adequate amounts of cobalt in a cow's rumen, the microbes cannot produce vitamin B12, leading to deficiency. In cases of severe deficiency, veterinarians inject cows with high doses of the vitamin. But it is more common to supplement cows with cobalt.

2. Protein Supplements

When forage quality is poor or insufficient for grass-fed animals, cattle may receive protein supplements. This supplemental feeding can consist of **soya bean meal, cottonseed meal or distiller's grains** (a byproduct of ethanol production). Protein is critical for muscle growth and weight gain, especially in young growing animals. But it is also required to produce the quality of meat that consumers want. With animals fed soya and corn mixtures, protein content is calculated based on the animals' needs, with any supplements mixed into the feed.

3. Energy Supplements

Cattle in feedlots are often given high-energy feeds like corn, barley or sorghum to promote rapid weight gain. In some cases, farmers add fats or oils to increase the energy density of the diet. These high-calorie diets make cows grow faster, getting them to slaughter weight quicker, allowing farmers to make more money. These foods are not ideal for cattle and can lead to digestive issues. One such illness is acidosis. Acidosis is an imbalance of microbes in the rumen. It causes cows discomfort, reduced feed intake and can lead to severe health problems. Bloating can result from excessive amounts of fermentable carbohydrates in the stomach. The gas accumulation and bloat are painful and have the potential to be fatal. Energy-dense diets can also contribute to lameness. Animals can become lame, where they can no longer walk, due to digestive imbalances causing laminitis.

4. Ionophores, Seaweed and Garlic

- **Ionophores**: Ionophores are feed additives that alter rumen fermentation to improve feed efficiency and reduce methane emissions through this process. Common examples include **monensin** and **lasalocid,** which help cattle convert feed into energy more efficiently and reduce the risk of digestive disorders like acidosis. Other additives that can be given to cows to reduce methane emissions are garlic and seaweed.

- **Seaweed**: Certain species of seaweed, particularly **Asparagopsis taxiformis**, are effective in reducing methane emissions in cattle. The active compounds in seaweed, such as **bromoform**, inhibit the activity of methane-producing microbes (methanogens) in the cow's rumen.

    Studies have shown that adding a small percentage of Asparagopsis to cow feed can reduce methane emissions by up to 90% and this would not impact digestion or milk production. Studies are not producing consistent results, though. This is because they are not carried out in real-life farming situations, so true efficacy is unknown. There are also safety risks around bromoform, which may have health and environmental impacts. Bromoform, classified as a possible carcinogen (Group 2B) by the **International Agency for Research on Cancer (IARC)**, is not considered safe. This is based on evidence that it can cause cancer in laboratory animals. It is also a toxic substance, for both humans and cattle, that can affect the liver and kidneys when consumed in high doses. Regardless, seaweed is an expensive additive to feed cattle, causing uptake to be low. There is also the problem of scaling seaweed production. Currently, there is no way to sustainably grow enough seaweed to supplement the **1 billion to 1.5 billion cows** being farmed at any given time.

- **Garlic**: **Garlic**: Garlic contains **organosulfur compounds** like **allicin**. These compounds have antimicrobial properties that target and reduce the activity of methanogenic microbes in the rumen. This is like seaweed but uses a different biochemical pathway. In cattle, garlic has been shown to reduce methane emissions by 20-40% in some studies, depending on the concentration used. But, like seaweed, adding garlic to the feed of cattle comes with hitches. Cows are not keen on the taste or smell of garlic; any garlic-based additives must be well blended with the rest of the feed. This blending is also important to ensure even garlic distribution. Overuse of the vegetable can cause digestive issues or changes in milk flavour for dairy cows. Garlic production would also have to be greatly increased for large-scale use in cattle farming, which is not possible at present.

As an alternative to feeding cattle digestion-altering compounds, masks have been developed to catch cow burps. A cow releases 95% of the methane it produces in burps. Methane-capturing masks, fitted over a cow's nose and mouth, trap or neutralise the methane. These masks can reduce methane emissions by up to **50-60%** in trials. But this device would be an extra cruelty added to the experience of being a farm animal.

5. Salt Blocks and Licks

Salt is often provided in the form of blocks or licks, sometimes mixed with minerals. These are essential for maintaining electrolyte balance and ensuring cattle consume adequate minerals.

6. Medications

In some cases, farmers add antibiotics or other medications to feed or water to prevent diseases in intensive farming systems. But antibiotics use faces greater scrutiny due to concerns about antibiotic resistance.

The Climate Change Connection

Beef farming, as with all forms of animal agriculture, contributes to climate change and is also highly vulnerable to its effects. The industry's reliance on natural resources, such as water and land, makes it susceptible to the impacts of a changing climate. The impact of beef farming is, at a slow rate, becoming more well-known. But there are still many who believe "blaming" beef for climate change and global warming is extreme. That it is either a government conspiracy, a ploy to make people buy electric vehicles, or scaremongering. To be clear, beef farming, or cattle farming in general, is not to blame for the climate crisis that we are facing. But it is a huge factor contributing to the critical situation that we are in.

Calculating CO_2 Equivalents

CO_2-equivalent (CO_2eq) is a standard unit used to compare the **global warming potential (GWP)** of different greenhouse gases. It expresses them as the amount of **carbon dioxide (CO_2)** that would have the same warming effect over a specific period (usually 100 years). This unit is a simplified way to measure and communicate the impact of greenhouse gases from different sources. Using CO_2eq, we can compare beef to 2 other metrics that are often used in the argument for changing how we live for the sake of the planet: aviation and car use.

When calculating the CO_2eq of different metric, we need to use averages as exact values differ across the globe. Factors of influence include local regulation and even the climate.

Cows

Key Variables:

1. Number of Cows Slaughtered Per Year:
    - Globally: ≈300 million cows.
2. Average Emissions per Cow:
    - Estimates for beef production range from **27 to 60 kg CO_2eq per kilogram of beef**, depending on farming practices.
    - Assuming, for simplicity, the average carcass yield per cow is around **300 kg** (yield varies by breed and farming methods).
3. CO_2eq per Cow:
    - Taking the mid-range emissions estimate (≈40 kg CO_2eq/kg of beef), each cow contributes: 40 kg CO_2eq × 300 kg of beef = 12,000 kg CO_2eq per cow.
4. Total CO_2eq:
    - Multiply the emissions per cow by the number of cows slaughtered: 300,000,000 cows × 12,000 kg CO_2eq.

Calculation

The total CO_2-equivalent (CO_2eq) emissions from global beef production are approximately **3.6 trillion kilograms (3.6 gigatonnes)** per year.

This estimate assumes:

- 300 million cows are slaughtered annually.
- Each cow produces ≈300 kg of beef.
- Each kilogram of beef generates an average of 40 kg CO_2eq.

Aviation

Key Variables

1. Global Annual Aviation Emissions:
    - Aviation contributes roughly **2.5% of global CO_2 emissions**, equating to about **915 million tonnes of CO_2** annually (as of recent estimates).
    - Including non-CO_2 effects (like contrails and high-altitude NOx emissions), the total **CO_2eq** impact is estimated to be about **2 – 3 times** higher, or **2.3 gigatonnes (2,300 million tonnes CO_2eq)** annually.
2. Passenger Kilometres Flown:
    - Total global aviation traffic is approximately **8.7 trillion revenue passenger kilometres (RPK)** per year.
3. Average Emissions per RPK:
    - On average, each passenger kilometre contributes **0.11 – 0.13 kg CO_2eq**, depending on aircraft efficiency, flight distances and fuel type.

Calculation Method

We can calculate aviation emissions by:

1. Using the total annual emissions figure (in this case, 2.3 gigatonnes CO_2eq).
2. Estimating per-flight emissions based on total passenger kilometres.

Aviation CO_2-Equivalent Emissions:

1. **Global Estimate (Including Non-CO_2 Effects):** Aviation contributes approximately **2.3 gigatonnes (2.3**

trillion kilograms) of CO₂eq annually, considering non-CO₂ impacts like contrails and high-altitude effects.
2. **Calculated Based on Passenger Kilometres:** Using 8.7 trillion passenger kilometres flown per year and an average emission of 0.12 kg CO₂eq per kilometre, the total is approximately **1.04 gigatonnes (1.04 trillion kilograms)** of CO₂eq.

The discrepancy in the totals arises from the first figure accounts for non-CO₂ effects, which can significantly amplify the climate impact of aviation.

Cars

Key Variables

1. Global Passenger Vehicles:
    - There are approximately **1.4 billion passenger vehicles** globally.
2. Annual Distance Driven:
    - The average passenger car drives about **15,000 kilometres (9,320 miles)** annually.
3. Average CO₂ Emissions per Kilometre:
    - Cars emit an average of **120 – 200 grams of CO₂ per kilometre**, depending on factors like fuel type, vehicle efficiency and driving conditions.
    - For simplicity, let's use an average of **150 g CO₂/km**.

4. Electric Vehicles (EVs):
    - EVs currently account for about **2% of the global fleet** and generally have lower lifecycle emissions, but for this calculation, we'll focus on internal combustion engine (ICE) vehicles.

Calculation

1. Total annual kilometres driven: 1.4 billion cars × 15,000 km
2. Total emissions: Total km driven × 150 g CO_2/km

The total CO_2 emissions from passenger cars globally are approximately **3.15 trillion kilograms (3.15 gigatonnes)** per year.

This estimate assumes:

- 1.4 billion cars.
- An average of 15,000 kilometres driven annually per car.
- 150 grams of CO_2 emitted per kilometre.

Data for these calculations was sourced from:

- The IPCC Fifth Assessment Report (AR5) Global Warming Potential (GWP) Data [177][178]
- United Nations Framework Convention on Climate Change (UNFCCC) [179]
- The International Energy Agency (IEA), Our World in Data (OWID) [180][181]
- The World Resources Institute (WRI), The International Air Transport Association (IATA) [182][183]
- ICCT (International Council on Clean Transportation) [184]
- FAO (Food and Agriculture Organization of the United Nations) [185]
- EPA (United States Environmental Protection Agency) [186]

To summarise:

- Beef farming results in **3.6 gigatonnes** CO_2eq annually
- Passenger aviation results in **1.04 gigatonnes** CO_2eq annually
- Car travel results in **3.15 gigatonnes** CO_2eq annually

It is clear to see that beef farming is a significant source of greenhouse gases, particularly methane and nitrous oxide. The large-scale production of these gases by the beef industry accelerates global warming. They are contributing to more frequent and severe climate events, such as droughts, floods, and heatwaves.

The carbon footprint of beef is also among the highest of all food products, making it a major contributor to the carbon footprint of the global food system. Reducing beef consumption is one of the most effective ways to lower an individual's carbon footprint and combat climate change. It is even more effective than taking all cars off the road, by our calculations above.

An extra problem with the relationship between beef farming and climate change is that the beef industry is highly vulnerable to its effects. This is seen particularly in regions where water resources are already limited. Temperature and rainfall changes can impact water availability for cattle and feed crops. This leads to reduced productivity and increased costs for farmers. As temperatures continue to go up, both crops and animals need more water.

Climate change can also exacerbate the spread of diseases and pests. Both of which affect the health of cattle and the productivity of crops. These challenges mean more resources are used, further increasing the environmental impact of beef farming. Water and feed consumption increase. Animals need more veterinary care.

The Beef Carbon Reduction Scheme (UK)

The **Beef Carbon Reduction Scheme (BCRS)** is a UK government initiative designed to reduce the environmental impact of beef production. The scheme aims to cut carbon intensity, which is the carbon emitted per unit of beef produced. It also encourages sustainable, climate-friendly practices in beef farming [187].

The main parts of the scheme are:

- Optimising feed efficiency
- Using methane mitigation technology
- Improving grazing management
- Efficient Manure Management
- Making genetic improvements

1. **Optimising Feed Efficiency:** The BCRS encourages farmers to optimise feed by improving feed efficiency. They are to feed cattle in a way that maximises growth and minimises waste. This can involve using more sustainable or locally sourced feed ingredients. They can use enhanced quality feed with additives that make digestion more efficient and cut methane emissions.
2. **Methane Mitigation Technologies:** The BCRS encourages farmers to explore methane-reducing technologies. These include the feed additives mentioned above. The seaweed and garlic supplements mentioned above are examples of this strategy. These technologies aim to make beef production more sustainable, reducing emissions throughout the animal's life.
3. **Improved Grazing Management:** How cattle graze and move across the land can significantly impact both **carbon sequestration** and emissions. The BCRS encourages farmers to adopt **rotational grazing practices**. They are to move animals between different pastures to allow grasslands to regenerate. This improves the ability of the soil and plants to absorb carbon. The practice also reduces **overgrazing**, land degradation and increased emissions.
4. **Efficient Manure Management:** The BCRS advocates for better manure management systems. These include storing manure in sealed environments or using it for **biogas production**. Biogas production can offset emissions by generating energy. Farmers can also apply manure more efficiently to fields to avoid excess nutrient runoff and nitrous oxide emissions.

5. **Genetic Improvements:** The scheme aims to breed cattle that are naturally more feed-efficient and produce fewer emissions. This could involve selecting animals that grow faster on less feed. These genetic improvements could reduce methane emissions by improving the efficiency of digestion.

The Role of Early Slaughter in Carbon Reduction

The early slaughter of animals also features in this scheme as one **carbon-reduction tool** among many. The logic behind it is that younger cattle tend to produce **less methane** since they spend less time in the **rumination** phase of their life. This occurs when their digestion shifts and methane production increases.

Financial Incentives and Monitoring

The BCRS uses financial incentives for farmers who adopt these carbon-reducing practices. The scheme also includes measurement tools to track carbon emissions and progress.

By encouraging farmers to adopt these practices and offering financial encouragement, the scheme aims to make beef production more **sustainable** and **climate-friendly**, while also helping the UK meet its carbon-reduction targets.

Sent Away for Slaughter

In some countries, such as China, France and Mexico, cattle are sent overseas for slaughter. The shipment of live cows overseas for slaughter is often referred to as the **live export industry**. Economic, logistical, and market considerations drive this controversial practice. It is cows that are most commonly transported by sea, although air transport is also used for high-value livestock or breeding animals [188].

These transports go to countries with high demand for fresh meat, with little local supply. Exporting live animals also allows these countries to maintain control over the slaughter process. This ensures the meat complies with local religious or cultural practices, such as **halal** or **kosher** slaughter.

Economic factors also play a significant role. For some beef-exporting nations, such as Australia or Brazil, selling live animals can be more profitable. It is cheaper to export live animals than process and export chilled or frozen meat, due to the costs of maintaining the cold chain.

The Logistics of Live Export

Cows are most often transported by sea in specialised **live export ships**. These vessels are equipped with pens, ventilation systems and feed storage. These are vital to support the animals during voyages, which can last from several days to weeks. In rare cases, cows are flown by cargo planes. But this only happens if the animals are high-value breeding stock or the importing country has urgent demand.

Before these stressful journeys, cows have health checks and vaccinations to meet the importing country's requirements. Onboard ships, conditions are crowded, unfamiliar and confusing for animals. Long travel times often lead to injuries, dehydration and even death during transit.

Animal Welfare and Environmental Concerns

The live export industry has faced widespread criticism about animal welfare. These long journeys, often in overcrowded conditions, lead to significant stress and suffering. The slaughter practices in some importing countries may not adhere to the humane standards in the exporting nations. Activists and animal welfare organisations have documented cases of mistreatment during transit and upon arrival, fuelling calls to ban or restrict live exports. Governments have already started to act on these calls, and bans have already been enacted in many countries.

The environmental impact of live export is also significant. Transporting animals by ship or plane adds to the greenhouse gas emissions of beef production. The ships used to transport cattle have very high energy costs. The use of huge amounts of fossil fuels for ventilation and water delivery systems, as well as moving the extremely heavy vehicles many hundreds of miles, produces large amounts of CO_2. The ventilation systems are particularly important to ensure animals don't suffocate. This is a particular problem on these vessels due to the nature of air movement at sea. As a ship moves, the air can move faster, slower or at the same speed as the ship. When air moves at the same speed as the ship, airflow onboard reduces. The lack of air movement means that the stale air from the animals does not move. This causes a build-up of foul air and increased CO_2 that can cause animals to suffocate. For this reason, all live export ships use huge industrial ventilation systems that have equally huge energy costs.

Live animal transport isn't only used to move beef cattle. Other animals such as sheep, pigs, and chickens are transported by the same methods for slaughter, breeding or fattening. Sheep, for example, are frequently transported live for international trade, such as exports from Australia to the Middle East. Not limited to meat production, live animals are also transported for milk production, wool harvesting and use in entertainment or research.

Tragedy in the Air

In October 2014, 174 sheep died on an export cargo flight from Perth to Singapore. The animals were part of a consignment of more than 2,000 animals flown by Singapore Airlines. Their deaths were documented to be the result of poor ventilation leading to asphyxiation [189].

Veterinarians Against Live Export investigated the incident and produced a report explaining how the animals were trapped until they expired. On the day of the flight, workers loaded 2200 sheep into 21 three-tiered wooden crates and 15 two-tiered wooden crates between 10:00 am and 3:30 pm. The floor area of each tier of the crates was 6.13 m^2. Twenty-three or twenty-four sheep went onto each tier, giving an average pen area of 0.242 m^2 per animal. During the 5-and-a-half-hour flight, data recorded by the plane's computers logged temperatures between 26°C and 35°C in areas where the sheep were. The final mortality investigation report stated, "temperatures inside the crates may have been higher than those recorded in the holds by the aircraft's temperature monitoring system."

"It's very upsetting to learn of the death of any animals, let alone 174 sheep on an aircraft," - Australian Livestock Exporters' Council chief executive Alison Penfold

In the seven years prior to this event, more than one million sheep, cattle and goats had been exported on planes in the previous seven years with low mortality rates reported.

"Between 2008 and 2013 there were 1,732 individual flights of live animals, being sheep, cattle and goats," - Alison Penfold

The total number of live exports up to the date of the incident was around 1.2 million animals, and of that, there had been 340 stated mortalities. For this reason, people believed that air freight was a low-risk method of transport. Comparing this line of thought with how this event would be perceived if the passengers were human and not sheep, it is easy to see why animal activists were mobilised to act against the practice.

At the time, Dr Bibba Jones, a representative of the RSPCA, said that they had grave concerns about the deaths and understood that the sheep had died of heat stress and not asphyxiation. "I mean, if you could imagine if that happened on a passenger plane, how much effort would be made to address the problem… Anything that has a death rate of 8 per cent is absolutely shocking for something like air transport, where you would expect a very, very low mortality rate," she said.

The Gulf Livestock 1 Disaster

On the 2nd of September 2020, the *Gulf Livestock 1*, a livestock carrier, sank 100 miles west of Amami Ōshima Island, a Japanese island. Caught in rough seas caused by Typhoon Maysak, the ship sent a distress signal at 1:40 am. But waves overwhelmed the vessel, which later capsized after the engines failed. It was carrying a cargo of 5,867 live cattle from New Zealand to China by sea [190].

The Japanese Coast Guard rescued one of the 43 crew members on board on the day of the accident. They found another, unresponsive, two days later. But both died shortly after rescue. The coast guard also rescued a third crew member on the afternoon of the 4th of September. There were several cattle carcasses in the same area, but they considered the entire cargo lost at sea.

This disaster led to New Zealand banning the live export of animals by sea in 2023. The transportation of live animals by air is still legal, but due to being more difficult to organise, it is rarely used.

Veganism and the Shift Away from Beef

Adopting a vegan lifestyle, excluding all animal products, not only beef, is a powerful way to reduce the environmental impacts of farming. Plant-based alternatives, such as lentils, beans and tofu, have much lower environmental footprints. They produce far fewer greenhouse gases. They can also provide similar nutritional benefits without the associated environmental costs.

Beef, as a food product, is inefficient in the extreme. Growing a whole cow to enable the harvesting of meat requires huge inputs for little return. For example, of the feed provided to cattle, only 4% of the protein and 3% of the total calories end up in the meat. The other 96% of protein, and 97% of calories, are lost in growing the cow and keeping it alive. Producing 1 kg of beef requires 25 kg of fodder and 15,000 litres of water. Going beef-free would save 2 billion hectares of land. Going vegan would save 3 billion hectares of land and remove roughly 800 billion tonnes of CO_2 from the atmosphere over 100 years. That's 8 billion tonnes of CO_2 removed from the air per year without making any other changes to our lives.

Every 1 and a half years, **more animals are killed than humans that have ever lived**. This is not a sustainable practice. Eating a plant-based diet reduces land use and, as noted above, removes carbon from the air, storing it in plants and soil. Harvests such as nuts, citrus fruits and olives grow from trees and store carbon. When these trees grow in the most suitable environments, their farming becomes reforestation.

Veganism addresses environmental concerns. But it also promotes ethical food systems that do not rely on the exploitation of animals. The beef industry causes many ethical issues. From the manipulation of bulls to harvest their semen. Keeping animals confined for their entire lives. There is feeding them an unnatural, high-calorie diet that can cause them pain or death. By choosing plant-based alternatives, individuals contribute to the development of food systems that respect animal welfare while promoting the well-being of all living beings.

Conclusion

Beef farming is one of the most environmentally damaging forms of agriculture. Contributing to climate change, deforestation, water pollution and biodiversity loss. The resource-intensive nature of beef makes it a major contributor to environmental degradation. This is through a combination of pollution, particularly water pollution and habitat destruction.

Plant-based alternatives to beef, including meat substitutes, need far less water, land and energy to produce. For example, producing a kilogram of lentils or beans requires a fraction of the water needed to produce the same amount of beef. It also generates far fewer greenhouse gas emissions. These alternatives avoid the pollution and resource depletion associated with beef farming. They are a more sustainable choice for consumers concerned about the environment.

Adopting a vegan lifestyle helps reduce the environmental impact of food. Individuals play a key role in this change through their diets and daily choices. This shift would help to mitigate climate change and protect natural resources. It would also support the development of more sustainable and ethical food systems. These systems allow land to grow food to feed people, reducing levels of food scarcity for many and feeding the world.

Chapter 7: Hogs and Harms

"When you go out there, it's not just a whimsical thing you start into... It's something you've thought out... When you see a hog farm, he didn't just decide overnight, 'I'm going to put me up a hog house.'" – Eva Ketelsleger, Pig farmer

An Introduction to Pig Farming

The public rarely considers pig farming, whether they consume its products or not, but humans have domesticated pigs for thousands of years. Pig farming began approximately 9,000 to 10,000 years ago with the domestication of wild boars. In China and the Near East (modern-day Turkey, Syria, and Iraq), people captured and farmed boars. Early explorers of the region brought this practice back to Europe during the Neolithic period, around 6,000 to 4,000 BCE. This occurred as part of a wider wave of agricultural expansion. People transported domesticated plants and animals into Europe, including dogs, goats, wheat, and legumes, domesticating them for commercial purposes. As early farmers migrated westward, they also brought domesticated pigs, derived from wild boars, as part of their farming practices.

Genetic studies have suggested that the boars farmers took with them were soon bred with local European boar species, producing the pigs farmed today. These pigs largely adapted to local environments and became distinct from their Near Eastern ancestors. It was these animals that were later taken to the Americas, where hog farming became a huge industry of its own.

Introduced by European explorers and settlers as a reliable source of food in the 1400s, the pigs were highly adaptable. They were quick to reproduce and able to find their food through foraging. It was during the **19th century** that the industrialisation of pig farming began. Fuelled by urbanisation and increased demand for pork, a shift toward pig farming on a larger scale took place. Farmers used selective breeding to develop specific traits, encouraging faster growth and higher fat content. By the **mid-20th century**, advances in genetics, veterinary science and nutrition developed into the modern industrial farming of today.

Large-scale intensive pig farms were developed in the mid-to-late 20th century. But the growing demand for pork prompted the creation of **concentrated animal feeding operations (CAFOs)**. As with the farming of cows and chickens, pig (or hog) CAFOs raise pigs in confined spaces. Farmers control the environment to maximise efficiency and minimise costs. The use of artificial insemination and genetic engineering is widespread. This allows farmers full authority in the farming of pigs, including the use of processed grains and medications.

From Farrow to Finish

The process of pig (or hog) farming is similar to that of cattle farming, using artificial methods. Breeders make boars produce sperm through manual methods. Farmers manipulate female pigs with hormones to enable fertilisation. Once pregnant, they remain confined, even after giving birth. Piglets have their teeth clipped and testicles pulled off before being fattened for slaughter.

Breeding

Pig farming starts with the use of **artificial insemination (AI)** for timed births and selective breeding. Although natural mating with boars (adult males) happens in smaller or traditional operations, AI is preferred in large-scale farms. This allows for control over genetics and the selection of specific traits. Features like fast growth, resistance to disease and desirable meat are bred into animals.

Boars are selected for their traits and are often part of a breeding program. These animals are trained to mount a dummy sow, allowing the handler to manually stimulate the penis. By applying pressure, mimicking the natural grip of a sow's cervix, they encourage ejaculation. Boar ejaculation occurs in three stages, or fractions: clear pre-sperm, sperm-rich and a gelatinous fraction (a thick plug). Breeders only collect the sperm-rich fraction. The handler, wearing a disposable glove coated with a lubricant to grip the penis, directs the ejaculate into a sterile collection container. In advanced facilities, an **artificial vagina (AV)** or collection funnel may be used, mirroring the method used in cow farming.

The sperm-rich fraction of semen is then separated from other fluids. It is filtered to remove debris and gelatinous material and analysed, with poor-quality samples discarded. The semen is then diluted with a specialised extender solution. The solution preserves the sperm and allows a single ejaculate to inseminate several sows. Semen samples are stored at approximately 16-18°C (60-64°F). The samples can also be frozen, though this is less common for boar semen as freezing reduces fertility compared to fresh samples.

Female pigs, either sows or gilts if this is to be their first litter, are selected for breeding when they are in oestrus. In pigs, ovulation occurs in the last 24–36 hours of this period. The use of a boar, or boar pheromones, often in the form of a spray, stimulates the standing reflex necessary for insemination. Farmers warm the semen sample to body temperature before insemination. A specialised **AI catheter** is then inserted into the vulva and then the cervix. Once in place, the semen is attached to the catheter and deposited into the pig. At this point, the sow, or gilt, rests for about 30 minutes to allow sperm to move up the reproductive tract. This procedure will be performed a second time when the pig is receptive, to maximise the possibility of success.

Once bred, a sow's gestation (pregnancy) lasts approximately **114 days**, during which time sows are often kept in **gestation crates** or group housing. Gestation crates are small metal enclosures about 2 x 6.6 feet, or 61 x 201 cm. They confine pregnant sows for the majority of their pregnancy. The sows can lie down, stand up, or sit, but cannot turn around or move freely.

Farrowing (Birth)

Pregnant sows move to **farrowing pens** or **crates** shortly before giving birth to their litter. Farrowing crates are like gestation crates. They restrict movement to prevent sows from crushing piglets. Each sow produces **8–12 piglets per litter**, depending on breed and genetics, and farrowing pens allow piglets to nurse, while preventing injury. Like gestation crates, sows cannot turn around or move freely. They cannot even move backwards or forwards, as the crates are as long as the sow. The only movement allowed is to lie down, sit or stand. In the United States, China and Brazil, gestation and farrowing crates are extensively used. Many other regions, including the European Union, Australia and Canada, have banned the use of this confinement.

Once born, piglets will stay with their mother for the first **3-4 weeks** and feed on her milk. But farmers may give supplemental **creep feed** to help piglets transition to solid food.

Weaning

At **21-28 days old**, piglets are **weaned**, or separated from their mother. These piglets move to a specialised facility called a **nursery** if they are not raised on a farrow-to-finish farm.

Weaning is a critical but stressful phase for piglets. They must adapt to a new environment and diet with no mother for guidance or support. Workers watch piglets closely during this time. If any illnesses, such as respiratory diseases and digestive issues, develop, treatment is given, or piglets are removed. These illnesses are more common during this period of a pig's life, particularly in these environments. When the piglets have moved, the sow returns to the breeding area to start a new reproductive and breeding cycle.

Nursery Stage

Weaned piglets, now called **feeder pigs**, spend **6-8 weeks** in the nursery. They remain in temperature-controlled pens and are provided with growth-promoting feed. Piglets are also vaccinated and monitored for diseases during this stage. While in the nursery, they grow from around **4.5-7 kg (10-15 pounds) at weaning** to **23-27 kg (50-60 pounds)** before transfer to the growing/finishing stage.

Growing/Finishing Stage

Feeder pigs leave the nursery and move to a **finishing farm** or **finishing barn**, depending on the operation. They remain there for **4-6 months**, gaining weight on a diet consisting mainly of **corn, soybeans and supplements**. During this phase, pigs grow from about **23 kg (60 pounds) to 113-136 kg (250-300 pounds)**. This is the standard weight for market-ready pigs. Climate-controlled finishing barns give pigs space to move, but these remain limited in industrial systems.

Once at market weight, pigs go to **slaughterhouses**. These facilities "process" animals into different products. This includes fresh cuts, such as pork chops, as well as cured meats like bacon and ham, and by-products like gelatine, lard, and fertiliser materials.

Specialised Systems

Within pig farming, there are various systems for getting piglets to slaughter weight pigs. These systems can involve selling pigs at different stages of their lives to live in facilities that specialise in growing pigs at each life phase, or keeping the pigs in one farm for the whole of their life.

- **Integrated Pig Production Systems** - An integrated system is a vertically managed operation where all stages of pig production occur within one facility. This includes breeding, farrowing, growing and finishing. It is often used in large-scale industrial pig farms, ensuring streamlined operations. The stages of an integrated system are standardised:
  - **Breeding and Gestation** - Sows or gilts are bred, either naturally or through artificial insemination. Pregnant sows remain in pens or gestation crates for about 114 days (the gestation period).
  - **Farrowing (Birth and Weaning Stage)** - Sows give birth (farrowing) in pens or farrowing crates, designed to protect piglets from being crushed. Mothers nurse their piglets for 2-4 weeks. During this stage, piglets often have their tail docked and teeth clipped to prevent injuries and infections.
  - **Nursery/Weaner Stage** - After weaning (around 3-4 weeks old), piglets move to nursery facilities, where they stay in groups. During this time, they feed on a specialised diet to promote growth while adapting to solid food.

- **Growing/Finishing Stage** - At around 8-12 weeks of age, pigs transfer to growing and finishing units, remaining there until they reach market weight (around 6-8 months old). These units focus on efficient feed conversion and fattening.

Non-Integrated Pig Production Systems

In **non-integrated systems**, separate farms or producers manage different stages of pig production. This allows farms to specialise in that life stage of the pig. This specialisation is more common in smaller or traditional farming setups. The stages within the non-integrated systems have the extra complication of disease spread. This can result from diseased pigs bringing illnesses to the new setting. The logistics of moving pigs from one farm to another are another consideration. Even so, the two systems are broadly the same as integrated systems:

1. **Farrow-to-Wean**: Farms that only handle breeding and weaning sell piglets to nurseries or grower farms.
2. **Wean-to-Finish**: These farms buy weaned piglets and raise them until they are ready for market.
3. **Grow-to-Finish**: Operations that buy nursery pigs and grow them to market weight.

Factory Farmed or Pasture Raised

CAFOs

Globally, around **98% of pigs** are raised in factory farms. The remaining roughly 2% are pasture-raised or farmed in small-scale or mixed systems. The efficiency of scale has made factory farming the dominating method for producing pork and pig products [191].

Approximately 98% of factory farms or **industrial-scale operations** are **Concentrated Animal Feeding Operations (CAFOs)**. In these facilities, thousands of animals live in controlled environments. Farmers can control food, water, temperature and even light schedules. Farmers breed pigs for rapid growth, feeding specialised fattening diets. These systems prioritise efficiency and profit. Cramped living conditions maximise the use of space while limiting mobility for the animals.

China, the United States and the European Union are the leading users of industrialised pig farming methods. China is the largest producer and consumer of pork, with a heavy reliance on industrial farms. The United States raises over 90% of its pigs in factory farms. In the European Union, particularly in Eastern Europe, intensive farming methods are heavily utilised.

Pasture-Raised Pigs

Raising pigs in pasture-based systems is more common in **New Zealand**, **Australia** and some parts of **Europe**, such as Denmark. In these countries, traditional or small-scale farming practices persist. Within the United States, niche producers, like those raising heritage pig breeds, also work in this way. Globally, a very small percentage of pigs in pasture or free-range systems can go outdoors. This time outside allows animals to roam, forage and exhibit some natural behaviours. The pigs, usually fed forage, will receive supplemental grains and kitchen or farm scraps. This practice is common in **organic** or **heritage** farming systems, and consumers value pasture-raised pork for its higher welfare standards and perceived superior flavour.

Other Farming Practices

The remaining number of pig farms is a variety of mixed or small-scale systems [192]. These include smallholder farms, where pigs live in backyards and feed on crop by-products. Extensive systems hold pigs in open pens or semi-confined spaces. Traditional free-range systems raise pigs semi-wild.

Teeth, Testicles and Tails

It is worth explaining the procedures that piglets go through as part of the farming process. These procedures - **tail docking**, **teeth clipping** and **testicle removal (castration)** - are common in conventional pig farming and large-scale operations. They're performed to prevent injuries, improve pig welfare (within the farming system's context) and enhance meat quality. But all three procedures are controversial due to concerns about animal welfare.

Tail Docking

Pigs, especially when confined, may bite each other's tails due to stress, boredom or frustration. This biting behaviour leads to injury, pain and infections. To prevent this, piglets have their tails docked within days of their birth. The docking removes part of the piglet's tail, usually using a heated cutter, sharp blade or rubber ring, leaving only a stub behind.

The procedure causes a lot of stress and pain for piglets and often happens without anaesthesia or pain relief. The practice also fails to address the cause of the tail-biting behaviour, the emotional distress that the animals live with, as it is cheaper and easier to remove the tail.

Teeth Clipping

Piglets are born with very sharp canine teeth. To prevent injury to the faces of siblings and teats of sows, farmers clip the teeth, making them blunt. Teeth are clipped using clippers or grinders, again with no pain relief.

Testicle Removal (Castration)

The testicles of piglets are surgically removed to reduce aggression and sexual behaviour in boars. This makes them easier to handle. The procedure also helps prevent boar taint, an unpleasant smell and taste in pork. Boar taint can happen in uncastrated male pigs due to hormones like androsterone and skatole building up in the meat. Typically, castration occurs manually within the first week of life. A scalpel or blade is often used, and usually, no anaesthesia is provided. The testicles are sensitive due to nerve endings, so the procedure causes significant pain and distress.

Pig Farming and Greenhouse Gas Emissions

Greenhouse Gases

Raising roughly 98% of pigs in CAFOs or industrial-scale farms produces an industrial-scale amount of waste. Like other forms of animal agriculture, pig farming contributes to greenhouse gas emissions. Animal agriculture alone produces 9% of greenhouse gas releases in the USA [193]. Producing pork and other pig products involves several stages, and each stage releases **carbon dioxide (CO_2), methane (CH_4)** and **nitrous oxide (N_2O)**. One of the primary sources of greenhouse gas emissions in pig farming is the management of manure [194]. Pigs produce large quantities of manure; one pig can produce as much manure as 6-8 humans. If not managed properly, this manure releases methane and nitrous oxide into the atmosphere. Where human waste is treated before being released, pig waste is not. This is an acute problem in regions such as Iowa, USA, where pigs outnumber people 7.6 to 1 [195].

On CAFOs, pig waste is collected in lagoons. The pigs live in confined areas with slatted floors that allow waste to fall into a pit below. From the pit, pumps move the waste into nearby lagoons. These lagoons are large enough that they can be seen in satellite images. Often a green/brown colour with a skin on top, they stand in stark relief from the surrounding environment. It is in these pits that methane is produced during anaerobic decomposition of manure. Manure releases nitrous oxide when farmers apply it to fields as fertiliser. Both gases are significant contributors to climate change due to their high global warming potential.

Energy Inputs

Heating, cooling and ventilation systems are necessary equipment for industrial farming. Pig farming is no different, requiring considerable energy inputs for daily running. Pigs need a comfortable temperature, roughly between 15-30°C, to maintain their health. Keeping the temperature cooler is a disease control measure, slowing the breeding of bacteria. It also prevents pigs from using energy to regulate their temperature. Heating nursery areas can involve using radiant heaters, heat lamps or under-floor heating. When temperatures are too high, farmers use fans, misters that spray pigs with water, or evaporative cooling systems. Ventilation fans remove stale air, reduce humidity, and can help lower the temperature. An important role for the fans is to prevent the build-up of gases. **Ammonia** and other gases are a danger to pigs and staff in confined areas. Running fans, sometimes 24 hours a day, reduces the risk by removing the air inside pig barns to the outside air.

As with poultry farming, lighting systems mimic day and night. Light cycles maintain the circadian rhythm of animals and promote growth. There are also a multitude of sensors that track the environment, temperature, humidity and air quality. The large amounts of energy used in these processes mostly come from fossil fuels, resulting in carbon dioxide emissions. Additionally, the transportation of feed, animals (particularly in non-integrated pig farming operations) and pork products contributes to the carbon footprint of pig farming.

The production of feed, principally soya and corn, is another source of greenhouse gas emissions. Cultivating these crops often involves the same deforestation and land-use changes used in beef and poultry farming. Natural environments are razed, releasing stored carbon dioxide into the atmosphere. Deforestation for feed production is a significant concern in regions like the Amazon. Large areas of rainforest, cleared to grow soya beans, contribute to climate change. The loss of biodiversity and destruction of critical ecosystems prevent future carbon sequestration.

Water Pollution and Scarcity

Pig farming is a major contributor to water pollution. This can exacerbate water scarcity in regions where water resources are already limited. Brazil, one of the largest pig meat producers, is already suffering from a lack of water because of pig farming. In Mexico, pig farming uses 4% of the country's freshwater supply each year [196]. In the United States, the total pork industry water use is 525,000,000,000 gallons per year. The water footprint accounts for 83-93% of the pork supply chain footprint, depending on the feed source [197]. The industry's impact on water quality and availability comes from both the inputs and outputs of its operations. Nutrient runoff and eutrophication are polluting groundwater and waterways. Pig farming uses huge fractions of freshwater, leaving little for consumption, household or industrial uses.

Nutrient Runoff and Eutrophication

The waste from industrial pig farms is mainly managed through manure spreading. This is a process known as **Comprehensive Nutrient Management**. Manure, collected from lagoons, is applied untreated to fields. Sprinkler systems spray the waste across fields. Trailing hoses, trailing shoes, or injection systems carry waste closer to the ground. Manure application is often carried out according to a lagoon's capacity and not necessarily the needs of the soil. This means that, to prevent overflow or starting a new lagoon, an amount of manure will be removed and applied across the soil and crops. This type of use focuses on the lagoon's needs, and not the soil's nutrient needs. As a result, over-applying manure is common. The ground becomes overloaded, and excess nutrients run off. Nearby bodies of water become polluted with nutrients, particularly nitrogen and phosphorus. This brings us back to **eutrophication**. Excess nutrients drive the overgrowth of algae. Oxygen levels drop, causing dead zones where aquatic life cannot survive [198].

Eutrophication is a significant environmental problem in many parts of the world, affecting rivers, lakes, and coastal areas. The dead zones created can have severe ecological consequences. They disrupt aquatic ecosystems and reduce biodiversity, even in regions that are not close to pig farms. This is because moving water will carry the waste and nutrients miles from where they first entered the water. The presence of pig waste in freshwater will also make the water unusable for people for recreation. Fishing, swimming and other water activities will stop as the water becomes too hazardous to enjoy.

> *The global water footprint of pigs in the period 1996-2005 was about 460 billion m3/yr, which was 19% of the total water footprint of animal production in the world (all farm animals) - (Mekonnen and Hoekstra, 2010, 2012)* [60] [61]

Water polluted with waste from pig farms is a serious threat to the health of waterways and the plants, animals and humans relying on them. Pig manure is particularly pathogenic, with more than 100 different pathogens excreted by pigs. The health risk that these pathogens pose should not be underestimated. 8.5 grams of manure can contain **hundreds of millions of bacteria** [198]. In the 1990s, there were a series of waste spills from manure lagoons. The damage resulting from the untreated waste flowing freely resulted in calls for a moratorium on new pig farming operations. Both environmentalists and politicians, including former U.S. Rep. Charlie Rose, called for change. One of the spills included the rupture of a waste lagoon at Oceanview Farms in Onslow County. This breach sent **25 million gallons of animal waste** into tributaries of the New River [199].

Water Consumption in Pig Farming

Pig farming is also water-intensive. It requires large amounts of water for drinking, cleaning and processing. Our World in Data states that 1,796 litres of fresh water are required for every kilogram of pork meat. This is considered an underestimate by other examinations. The Water Footprint Network has found the water footprint of **1 kg of pork** is in the range of **4,000-6,000 litres per kilogram.** This amount depends on the production system and feed composition [60] [61]. The volume increases to **66,867 litres** when weighted against water scarcity [200]. **Scarcity-weighted** water use represents freshwater use weighted by local water scarcity. In regions where freshwater is not in regular or large supply, this becomes a critical metric.

In these regions, the high water demands of pig farming exacerbate water scarcity, impacting both local communities and ecosystems. This is a major concern in areas where pig farming is experiencing rapid expansion.

Much of the water used in pig farming is embedded in growing feed, with corn and soya being the main ingredients of what pigs eat. As they grow, farmers feed piglets and juveniles a controlled amount of food. When they approach slaughter weight, pigs feed freely. This is to encourage them to gain as much weight as possible to maximise profits. Pigs are also selectively bred to reach slaughter weight as quickly as possible. They grow faster and heavier with less feed while producing less manure. But it is still estimated that it requires 6.4 kg of feed to grow 1 kg of pig meat [201].

The amount of water required to grow 1 kg of pig feed varies depending on the type of feed used. Typically, a mix of grains like corn, soybeans, wheat and cereals ground together, each component has a different water footprint. Below is an approximate breakdown:

Water Requirements for Common Pig Feed Crops

1. **Corn (Maize)**: A staple in pig feed, often making up a large portion of the diet.
    - Water footprint: 900-1,400 litres per kg
2. **Soybeans**: A primary source of protein in pig feed.
    - Water footprint: 1,500-2,000 litres per kg
3. **Wheat**: Another common component, often used as an energy source.
    - Water footprint: 1,300-1,800 litres per kg
4. **Barley**: Used in some regions as an alternative to corn.
    - Water footprint: 1,300-1,400 litres per kg

Blended Feed Average

In a typical pig feed mix, these crops are combined in varying proportions according to requirements. The blended water footprint of pig feed is around **1,200-1,800 litres of water per kg**, with a feed-to-meat conversion ratio of 6.4 to 1.

Another use of large amounts of water is keeping pigs cool, with many thousands of litres of water being used in spray and sprinkler systems running above pig enclosures. This water will run off pigs and through the slatted floors, adding to the manure collected beneath and pumped into lagoons.

Pigs	Amount of drinking water (litres/day)
Lactating sow	18 - 23
Gestating sow/boar	13 - 18
Fattening pig	3 - 10
Weaner	1 - 3

The Department of Agriculture, Environment and Rural Affairs [200]

Table 1

Contamination of Water Sources

Water pollutants from pig farms come from manure and wastewater resulting from biosecurity and cleaning. Manure entering water streams is common. Overloading soil with manure and spraying waste onto porous ground, such as limestone, makes contamination of freshwater inevitable. The sheer volume of this waste makes any environmentally friendly controls impossible.

Wastewater from farms is also a considerable pollutant. This water needs to be captured and treated before entering municipal waste systems. The improper management of waste contaminates water sources with pathogens, hormones and antibiotics. Other pollutants, including cleaning chemicals containing biocides, are also toxic to aquatic environments. These contaminants entering groundwater and surface water are a risk to human health and the environment.

Contamination of water supplies has serious consequences for local communities. Those who rely on well water or other untreated sources are at high risk of illness. Pathogens and other harmful substances in drinking water can cause a range of health problems. Gastrointestinal illnesses that mimic food poisoning and antibiotic resistance are both consequences.

Cleaning and Slaughter

Cleaning of pig barns begins after the removal of pigs. Each farming stage, farrowing, weaning, nursery and finishing takes pigs from one area to another. This means that pens are thoroughly cleaned and disinfected once empty. Pig farms are highly pathogenic spaces. Both humans and animals contract and spread contagious diseases as a result of pig farming.

The cleaning process has several steps. First, there is the removal of all dry material, including spilt feed or manure, which removes dust and debris. The area is then jet-washed and treated with chemicals, like hydrofoaming degreasers or alkaline detergents. Depending on set protocols, pens are jet-washed a second time before having disinfectant applied to all surfaces. Disinfectants must dry before pigs enter the pen, to ensure that all pathogens are killed. A CAFO housing **10,000 pigs** might use **130,000-290,000 litres of water,** as roughly 2000 litres of water are used per sow [202].

At the slaughterhouse, pigs are unloaded into pens. The vehicle used will be thoroughly cleaned before the next use to prevent disease spread. This can require 750 gallons of water per truck [197]. The slaughter process for pigs involves several steps, with water playing a vital role throughout. It is needed for maintaining the required animal welfare standards and hygiene. But it is also used as part of slaughter operations. Upon arrival at the slaughterhouse, pigs in the holding pens have access to drinking water to reduce stress. This water is often the first drink that pigs have after leaving the farm. On the journey to the slaughterhouse, which can take over a day, pigs have no access to water or feed. Reducing the stress of animals is also crucial for meat production. People consider the meat of stressed animals to be of lower quality.

Slaughterhouse workers render pigs unconscious before killing them. Some methods use electric stunning or bolt guns. But gassing is the most common technique for stunning, using carbon dioxide (CO_2). In the UK, 90% of pigs are rendered unconscious using CO_2 [203].

The process is not pleasant. Handlers move pigs into cages, which they lower into gas chambers. A valve pumps CO_2 into the chamber, with pigs remaining there until they are unconscious. Water is indirectly involved in this stage, as it is often used in cooling systems for electrical equipment or to clean the stunning area. Injuries to pigs are often observed as animals thrash around as they suffocate. These violent movements are the result of pain and panic that the animals feel as the CO_2 chokes them. Once unconscious, workers hoist pigs and cut their throats to sever major blood vessels. This is the same process of **exsanguination** carried out on cows and poultry. Each pig contains 3-5 litres of blood. If this is not collected and used to make a product, such as fertiliser or blood sausage, water washes the blood away, keeping the slaughter area clean.

After bleeding, pigs go into a scalding tank, filled with hot water, to loosen their hair and remove dirt. The bodies are then dehaired, using machines with rotating paddles to scrape off the loosened hair. In some facilities, hot water showers replace scalding tanks. This step requires significant amounts of water for both scalding and rinsing the carcass after dehairing.

The next stage is evisceration. This is the removal of the internal organs of the pigs. The carcass is then inspected for diseases or abnormalities. Water is crucial during this phase for cleaning tools and frequent rinsing of the work area to prevent contamination. Following inspection, workers split, trim, clean, and chill the carcass. High-pressure water spray cleans the carcass thoroughly, removing blood, bone fragments and debris. Water is also used to sanitise cutting tools and work surfaces.

Once cleaned, carcasses move to cold storage for rapid cooling. After the slaughter process, the whole facility is thoroughly cleaned and sanitised to meet food safety regulations. This involves using water mixed with detergents and disinfectants to wash floors, walls and equipment.

Slaughterhouse workers divide a pig's body according to how they will use it, but generally separate it as follows:

From a 120 kg pig:

- **Meat & Fat (Edible)**: ≈ 65-85 kg
- **Blood**: ≈ 3-5 litres
- **Organs/Offal**: ≈ 15-25 kg
- **Bones**: ≈ 10-18 kg
- **Skin**: ≈ 8-12 kg
- **True Waste**: ≈ 3-6 kg

Pig slaughterhouses are significant water users. Their processes use **1,000 to 2,000 gallons of water per animal,** depending on the facility's efficiency. This generates large volumes of wastewater. The waste of slaughterhouses, blood, fat, manure and cleaning chemicals can all sully this water. If not properly captured and treated, these contaminants can pollute nearby water sources. Some modern facilities incorporate water recycling systems in their drains. These processes reuse water, reducing consumption and the environmental impact. But water usage in pig slaughtering remains a concern, particularly in areas of water scarcity or poor wastewater treatment.

Air Pollution and Health Impacts – Nutrient Management

Pig farming is a significant contributor to planetary pollution. It adds to global air pollution primarily through the emission of greenhouse gases (GHGs), including **methane (CH_4), nitrous oxide (N_2O)** and **carbon dioxide (CO_2).** The carbon footprint of 1 kg of pork products is equal to **2.25 to 4.52 kg CO_2eq**. The number of different gases released from this form of agriculture totals **over 200**. Many of these are hazardous to human and animal health. This is particularly a problem for the workers on pig farms. Daily, they face exposure to these toxic gases. Outside the farm, communities are also endangered. Rates of respiratory and gastrointestinal illnesses, and several cancers, are far higher in these neighbourhoods. The rates of these illnesses also increase with the number of pigs being farmed.

Gas emissions arise from different stages of pork production. These include the enteric digestive processes of pigs and the resulting manure. But the primary contributor is the production of pig feed. The Food and Agriculture Organization (FAO) state that livestock production accounts for approximately 14.5% of all human-induced greenhouse gas emissions. This equates to about **7.1 gigatonnes of CO_2-equivalent** per year [185]. This figure encompasses emissions from various sources. They include enteric fermentation in ruminants, manure management, feed production and associated land-use changes. Within this sector, pork production is the second-largest contributor, following beef. CO_2 contributes approximately 81% of total emissions, CH_4 accounts for 17%, and N_2O makes up about 2%.

Breaking down the emissions specific to pig farming:

- **Feed Production**: This stage causes approximately 60% of the emissions in the global pig supply chain.
- **Manure Management**: Manure storage and processing contribute about 27% of the emissions.
- **Other Processes**: The remaining 13% of emissions result from various other processes within pig farming.

The emissions from pig farms follow the same patterns as those from the farming of other animals. Ammonia, methane and particulate matter are common results. These pollutants contribute to air quality issues that affect nearby communities and the broader environment. The impact that this can have, especially in areas with a concentration of pig CAFOs, is great. People change the way that they live, such as not using outdoor spaces or opening windows. Families move away from farms, abandoning their homes. Others, whose children have moved away to study, have encouraged them not to return. Those who return find that illnesses, once resolved, come back. There have also been many cases of residents taking farmers and meat processing companies to court. Having reached their limits, they use the court to ask for the farming to stop. They also look for financial compensation for the harm they have suffered.

Ammonia Emissions

Pig farms are a major source of ammonia emissions, most of which are discharged into the air during the decomposition of manure. The pungent gas contributes to the formation of fine particulate matter ($PM_{2.5}$) in the atmosphere. The gas is harmful to human health, particularly to the moist pathways in the respiratory tract (e.g., the mucous membranes of the nose, throat, and lungs). Ammonia (NH_3) is highly alkaline, with a pH of 11-12. When dissolved in water, it forms ammonium hydroxide (NH_4OH). This high pH makes it caustic. It can irritate and damage tissues by disrupting cellular structures and causing chemical burns.

Exposure to high levels of ammonia and particulate matter results in respiratory problems. Vulnerable populations such as children, the elderly and those with pre-existing health conditions are most at risk. At low concentrations, ammonia causes mild irritation, such as coughing, burning in the nose and throat and watery eyes. At high concentrations, the caustic effect becomes severe. It has the potential to lead to chemical burns, airway swelling and respiratory distress. In extreme cases, prolonged exposure can result in permanent lung damage.

Ammonia releases can also lead to acidification of soils and water bodies, harming plant and aquatic life. Ammonia (NH_3) in the air reacts with acidic compounds in the atmosphere. These pollutants, like **sulphur dioxide (SO_2) and nitrogen oxides (NO_x)**, come from industrial processes and vehicle emissions. Combined with ammonia, they will form compounds such as **sulfuric acid (H_2SO_4)** and **nitric acid (HNO_3)**. These ammonium salts are the fine particulate matter. They remain in the air for a finite amount of time before eventually settling on the ground. From here, they can dissolve in water, releasing the acids, sulfuric and nitric acid, into soils and waterways. This re-acidification contributes to acid rain and its harmful effects on ecosystems. This interplay between compounds of different sources released into the air shows that human actions are damaging to the planet. It isn't simply that each process creates pollutants. These pollutants can react, forming more reactive and dangerous chemicals. This is a cascade of harms.

Odour Pollution

The air pollution resulting from pig farming is more than the gases released into the air from manure and barns. The odour from pig farms is a significant issue for many communities close to these operations. The strong smell of manure and other waste products can make living conditions unbearable. The suffering that people endure has led to conflicts between farmers and residents.

Odour pollution can also have psychological effects. It contributes to stress and reduces the quality of life for those living near pig farms. In some cases, the odours from pig farms can be so strong that they affect property values. People end up unable to leave; they cannot sell their home, because no one wants to live near a pig farm.

North Carolina Hog Nuisance Lawsuits (2018–2019):

Background: Residents in eastern North Carolina filed multiple lawsuits against Smithfield Foods' subsidiary, Murphy-Brown LLC, alleging that the company's pig farming operations caused significant nuisances, including foul odours, pests and noise, which interfered with their ability to enjoy their properties.

Outcome: In five separate trials, juries awarded plaintiffs nearly $550 million in total damages. These verdicts held Smithfield liable for creating nuisances that adversely affected the residents' quality of life. - Yale Law School

Missouri Premium Standard Farms Case (2010):

Background: Thirteen plaintiffs living near a Premium Standard Farms facility in Gentry County, Missouri, sued the company over issues related to odours and pollution from the hog operations.

Outcome: A jury awarded each of the 13 plaintiffs $825,000, totalling nearly $11 million. The verdict acknowledged the negative impact of the farm's operations on the residents' ability to enjoy their properties. - Seeger Weiss LLP

North Carolina Nuisance Lawsuit (2018):

Background: Ten plaintiffs living near a Smithfield Foods hog farm in Bladen County, North Carolina, filed a lawsuit citing nuisances such as odours and noise.

Outcome: A jury awarded the plaintiffs $50 million in damages. However, due to state laws capping punitive damages, the award was later reduced to $3.25 million. – Justia.com

Ag-Gag Laws

Ag-Gag laws are legal measures designed to protect the agriculture industry. They prevent or deter whistleblowers, journalists and activists from exposing what many consider the truth. Unethical or illegal practices, particularly in factory farming, are to remain a trade secret. The implementation of these laws in 1990 originated in Kansas, USA. Activists from The Animal Liberation Front targeted fur farms. The exposure of what they found shook the industry. In response, the industry worked with politicians and lawmakers to form the Ag-Gag laws. Members of The Animal Liberation Front, founded in the UK, had faced prosecution under existing UK laws. But in the USA, the industry wanted specific laws targeting animal activists. The term "Ag-Gag" originates from "agriculture" and the word "gag". These laws silence critics by criminalising undercover investigations. They also prevent the dissemination of information about agricultural operations. Ag-Gag laws include provisions that:

1. **Prohibit Undercover Investigations**: Criminalise entering agricultural facilities under false pretences. These include misrepresenting yourself during a job application. People do this to gain entry and document practices inside the facility.

2. **Ban Recording**: Make it illegal to record video or audio without the owner's consent. Even doing so in public-facing areas of the facility is illegal.
3. **Mandate Immediate Reporting**: Individuals must report suspected animal cruelty within a short time frame. People cannot collect evidence. This often makes thorough investigations impossible.
4. **Penalise Trespassing**: There are enhanced penalties for trespassing on private agricultural property. These penalties are enforced even when the intent is to expose harmful practices.

These laws make uncovering the truth about animal agriculture much more difficult. Those caught, whether working under false pretences, taking photos, filming on farms, or tapping into CCTV systems, face long prison sentences. These sentences can be for more than 10 years.

"Criminalising whistleblowing shows an absolute contempt for concerned consumers and a fundamental insecurity over the product and methods used." – Former Iowa Representative Liz Bennet

Not every country has Ag-Gag laws, as breaking and entering or trespassing laws are often enough. But, in countries that do, the legal boundaries can be the same as those of terrorism. The work of those going undercover is important for the welfare of animals, but also for people who eat meat. In 2008, an investigation by the Humane Society produced footage of workers hitting cows that could not walk. The release of the footage led to an investigation and a recall of beef. The meat of cows from this farm was unsafe, as cows being unable to walk is a sign of high risk of disease and illness. The recorded employees faced charges of animal cruelty. Those against animal agriculture consider these laws a silencing tactic. That they are evidence that the animal agriculture industry wants to hide their practices, rather than improve them. Whether improvements would be for animal welfare, staff safety or the protection of the environment.

Ag-Gag laws are highly controversial. They impose on the rights of animals not to be harmed, and for staff to be safe at work. They prevent workers from whistleblowing. They also stop consumers from knowing where their food has come from. Advocates argue they are necessary to prevent biosecurity risks, trespassing and economic sabotage. But opponents view them as an affront to transparency, ethical accountability and free speech. They also highlight their role in concealing widespread animal cruelty and environmental harm. By restricting investigations and public scrutiny, Ag-Gag laws remain a controversial issue at the junction between agriculture, ethics and legal rights.

George Steinmetz

"In many cases, the food industry goes to significant lengths to prevent us from seeing how our food is produced. Access to this information is central to the personal decisions we make about what we eat, which cumulatively have huge environmental impact. This project seeks to show how our food is produced, so that we can make more informed decisions." – George Steinmetz

George Steinmetz is an award-winning New Jersey-based aerial photographer. Arrested on 28th June 2013, after taking pictures of a feedlot outside Garden City, he faced charges of breaking Ag-Gag laws. Steinmetz was taking photos from a paraglider, as part of his *Feed the Planet* project [204]. The reason for the arrest was initially stated by Finney County Sheriff Kevin Bascue as trespassing.

Steinmetz and his paraglider instructor, Wei Zhang, had parked on farm property. They took off from the property without permission and did not tell anyone that they intended to take photos. This reason was later changed to infringing on the right of the farmer to operate in peace. Many believed this to be a use of Kansas's Ag-Gag laws, the Farm Animal and Field Crop and Research Facilities Protection Act. This was the U.S.'s first Ag-Gag law, signed into law in 1990. After being held for a short period at Finney County jail, both Steinmetz and Zhang were released on a $270 bond.

North Carolina

North Carolina is the **second-largest pork-producing state** in the U.S., after Iowa. The state has over **2,000 pig farms**, housing nearly **10 million pigs**. Most of these farms are in the eastern counties of the state. This concentration of pigs also means a concentration of waste and pollution. The waste from **10 million pigs is equal to the daily faecal output of roughly 280 million humans** [205] [206]. Duplin County and Sampson County have some of the highest pig-to-human ratios in the U.S., stated to be 29 to 1 [207]. The pork industry in this region contributes billions of dollars each year to the state's economy. It creates thousands of jobs in farming, processing and related trades. The financial benefits make lawmakers reluctant to enforce regulations on the industry.

A 2014 report by the Environment North Carolina Policy & Research Center examined the waste discharges of agriculture in the region. It revealed that, in 2012, corporate agribusiness facilities, including slaughterhouses, were responsible for approximately one-third of all direct nitrate discharges into the state's waterways [208]. These releases of nitrate compounds represented almost 90 per cent of toxic water discharges. One of the larger concerns with this finding was that nitrates, being toxic, are particularly harmful to infants. Babies and young children who drink formula made with water high in nitrates are at risk from this pollution. Methemoglobinemia, or "blue baby" syndrome, is a disease that reduces the ability of blood to carry oxygen throughout the body. Consuming large amounts of nitrates, such as those found in polluted water, can result in this fatal disease. Chapter 3 – Reducing Water Pollution Through Veganism discusses this in more depth. This isn't an unknown danger. Nitrates are also linked, in studies, to organ damage in adults. The Environmental Working Group (EWG) published a report in 1996, highlighting the dangers of nitrates in water [209].

> *According to the EPA, industrial pollution has left more than 17,000 miles of rivers and about 210,000 acres of lakes, ponds or reservoirs, unable to support drinking, swimming, fishing or other uses. - The Environment North Carolina Policy & Research Center* [208]

Nitrates convert to N-Nitroso compounds in the stomach; these compounds are human carcinogens. There is now a large body of work, research, literature and studies showing the effects of consuming nitrate-polluted water. They have all found that drinking water with high nitrate levels can increase cancer risks [210] [211]. Studies on rats have looked at the effect of nitrate compounds in early life. They found that exposure to N-nitroso diethylamine as infants has cancer rates six times higher than those exposed after weaning [212]. Human epidemiology studies have also suggested that the risk to children may be similar. That cancer risks may be higher in those exposed to nitrate-contaminated water in the first ten years of life [213]. Danish research has linked nitrate-contaminated water to thyroid enlargement [214]. Several studies have found a possible link between nitrate and N-nitroso compounds and birth defects. This was first observed in animal studies, but has since been observed in human epidemiological studies [215] [216] [217].

Floyd and Florence

In 1999, Hurricane Floyd caused the release of vast amounts of pig waste into streets as the Category 2 hurricane brought about catastrophic flooding, affecting rivers and manure lagoons [218].

Hurricane Floyd made landfall on September 16th, 1999, near Cape Fear, North Carolina, bringing over 20 inches of rain to some regions and causing rivers like the Tar and Neuse to overflow. The flooding was unprecedented and affected vast regions of eastern North Carolina. The massive flooding caused over 50 hog waste lagoons to breach, overflow, or submerge entirely, releasing millions of gallons of liquefied manure to spill into rivers, streams and the surrounding environment. The floodwaters also killed an estimated 30,000 hogs, along with millions of chickens and turkeys, further adding to the pollution.

The waste that spilled into waterways led to high levels of nitrogen, phosphorus and pathogens like E. coli, creating a public health crisis and devastating aquatic ecosystems. The influx of nutrients later fuelled algal blooms, which led to oxygen depletion and massive fish kills in affected rivers and estuaries. Drinking water was also contaminated as the waste could not be contained.

The disaster drew attention to the environmental risks posed by CAFOs, particularly their waste management systems, beyond the pollution that people already lived with and, in response, North Carolina implemented a temporary moratorium on new hog CAFOs. There was also an exploration into alternatives to lagoon-based waste management. But progress in implementing widespread reforms has been slow, and the practice of pumping pig waste into lagoons continues.

> *"at least six swine lagoons have suffered structural damage, and 30 have reported discharges. At least three of those with structural damage have been breached – the worst-case scenario, in which walls collapse. One released 2.2 million gallons in Duplin County"* – Michael Graff, The Guardian

Hurricane Florence struck North Carolina in September 2018. The Category 1 hurricane was another devastating storm that highlighted the environmental and public health risks associated with industrial agriculture, particularly pig waste pollution.

Hurricane Florence made landfall on September 14th, 2018, near Wrightsville Beach, North Carolina and, being slow-moving, it brought ruinous rainfall and flooding; some areas received over 89 centimetres of rain. The flooding inundated rivers, roads, homes and agricultural facilities across eastern North Carolina.

> *"This increasingly severe, potentially unprecedented storm, is hurdling to the epicentre of animal agriculture in North Carolina... Because waste is managed using archaic practices, it presents a significant threat to water quality, primarily through runoff and/or breach or inundation of hog lagoons."* - Will Hendrick, The Waterkeeper Alliance [219]

Thousands of CAFOs are located in North Carolina's coastal plain, one of the severely affected areas. These facilities house millions of pigs, with their waste stored in open-air lagoons with no adequate flood protection. The hurricane caused at least 50 of the pig waste lagoons to rupture, overflow, or become overwhelmed, releasing millions of gallons of untreated pig waste into surrounding land, waterways and the environment.

The storm also killed over 5,000 hogs and more than 3.4 million chickens and turkeys. Their carcasses, combined with waste, contributed to pollution and disposal challenges. Bodies were floating freely in the streets, were tangled in fences and were even found on the doorsteps of homes.

Floodwaters mixed with animal waste, dead bodies and pollutants from lagoons spread pathogens, nitrogen, phosphorus and toxic substances into rivers and streams. Runoff from fields added to the pollution and fuelled algal blooms, depleting oxygen in water and causing fish kills. Flooded lagoons risked leaching pollutants into groundwater, endangering private wells that many rural residents rely on. Health issues that could not be mitigated by actions such as boiling were reported by residents. They included gastrointestinal illnesses and respiratory problems, and were linked to water contamination.

The storm caused billions of dollars in agricultural losses, including damage to CAFO infrastructure, livestock and crops, causing previous debates around the placement and regulation of CAFOs in flood-prone areas to be reignited. Critics argued for stricter oversight and the adoption of more sustainable waste management systems. Some advocates pushed for covered or closed waste systems, or technologies to process waste into energy or fertiliser. Such systems are extremely costly and beyond the reach of many farmers, meaning that uptake has been minimal. Environmental groups and affected communities called for more accountability from the livestock industry and stronger protections against future disasters, yet no actions of note have taken place. The communities most affected by

CAFO pollution were predominantly low-income and communities of colour, sparking discussions about environmental justice and systemic inequality.

Iowa

"What's been kinda surprising in Iowa, is really, the smell of death." - Patrick Hansen, Iowa resident

Iowa is the largest producer of pork in the United States, accounting for about 30% of the nation's total pork production. The state is home to about **24 million pigs** and, on average, produces over **50 million pigs annually**, raising and slaughtering pigs in constant cycles [220]. Statewide, pigs outnumber humans 7.6 to 1, but in the most densely farmed areas, such as Lyon County, pigs outnumber humans 60 to 1 [221] [222]. In 2024, the USDA published a census on agriculture in the United States that found that Iowa's pig farms produce over **13 billion gallons of liquid manure annually** [223].

The space taken up by pigs is relatively small, considering the number of pigs within the state, as they are raised in concentrated numbers in confined spaces. But the land used to grow feed crops for these pigs is vast. **23 million acres of farmland** in Iowa are dedicated to corn and soybeans, the primary feed for pigs. If this land were used to grow crops for humans, it could feed as many as 400 million people a year [224].

In Iowa, **ammonia, methane and hydrogen sulphide** emissions from manure lagoons and barns are common, and studies have found higher rates of **asthma** and other **respiratory problems** in the communities near CAFOs [225]. The persistent odours, from manure spraying and pig barns, make outdoor activities unbearable for some residents, and property values drop the closer they are to CAFOs, due to odours, pollution and increased traffic from farm operations. Local economies in these communities often see little benefit, as many CAFOs are owned by large corporations, not local farmers.

> *"The supporters of [CAFOs] like to say it's the future of food and farming and I think that's nonsense. There are people who wanna farm in a way that supports the land and are doing it, and a lot of them are out here – it's one of the reasons we moved out here. So, to say that that's the future is, to me, just buying into propaganda and messaging from the company. If we invested the amount of time and energy we do into conventional agriculture and subsidies into small-scale, localised agriculture, that support family farms, we would not be having these problems, we would not be having these fights."* Kristy Allen, beekeeper and Burrnett County Resident

Iowa's drinking water is also severely affected by nitrate runoff from manure and fertilisers. In some areas, nitrate levels in water exceed the EPA's safe limit of 10 ppm, a level that is still higher than that of many countries. In rural areas of the U.S., it is common for people to rely on wells as a source of freshwater, and Iowa is no exception, but wells are particularly vulnerable to water pollution resulting from the runoff of fertilisers. Many wells tested in Iowa have been shown to have nitrate levels high enough to pose health risks, including blue baby syndrome and cancer.

Between 2013 and 2023, 179 manure spills were reported in Iowa, with stated volumes released ranging from **500 to 1 million gallons.** Many of these spills were linked to pig farms and the polluting of local waterways, such as the Raccoon and Des Moines Rivers [226]. These spills kill fish, degrade water quality and make rivers unusable for drinking water or recreation. The problem is so persistent that the Des Moines Water Works spends **millions of dollars annually** to remove nitrates from the water supply [227]. The utility company once sued three upstream counties for nitrate pollution, highlighting the burden on urban areas caused by agricultural practices. Iowa is also a leading contributor to the **Gulf of Mexico's dead zone** due to the amount of nutrient runoff, primarily from agricultural practices tied to pig farming.

Within farms, the situation is even more toxic. Workers in Iowa's pig farms face high exposure to **ammonia, hydrogen sulphide and particulate matter**, leading to chronic bronchitis, asthma, respiratory distress and an increased risk of contracting zoonotic diseases, such as **MRSA** and **swine flu** [228] [229]. Slaughterhouse workers are often immigrants or low-income individuals and must endure repetitive motion injuries and poor working conditions while earning low wages with limited protections.

Iowa's pig farming industry, while economically significant, has caused severe environmental and health challenges, particularly for rural communities. From nitrate-contaminated drinking water to air pollution and declining property values, the consequences of industrial pig farming in Iowa are profound. Advocates have called for a **moratorium on new CAFOs,** and they continue to push for sustainable practices and stronger regulations to address these issues. Iowa's government has faced criticism for lenient regulations on CAFOs, while the **Iowa Department of Natural Resources (DNR)** handles monitoring CAFOs, but lacks resources for strict enforcement.

Burnett County St. Croix

In June 2019, Jim Melin, the former town chairman of Trade Lake in Burnett County, played a central role in a controversy surrounding a proposed large-scale pig farm in the area. Melin sold some of his land to Cumberland LLC, the developer of a planned $20 million CAFO, which was intended to house up to 26,350 pigs. It would have been the largest pig breeding facility in Wisconsin [230].

Melin, who approached Cumberland LLC to sell his land, was accused by community members of having a conflict of interest. This was because he conducted land sales with Cumberland LLC while serving as the town chairman. Critics argued that his actions undermined trust and raised ethical concerns about his role in facilitating the project. Residents were so angered that they sued to remove him from office, alleging that he violated state ethics rules.

> *"It smelled like pigs… Pig farms smell." - Trade Lake, Iowa resident Howard Pahl*

Residents and environmental groups strongly opposed the project. This was due to concerns about groundwater contamination, air quality issues from ammonia and hydrogen sulphide emissions. They were also worried about the potential for large-scale manure spills. Burnett County relies on wells for drinking water. The risk of water pollution heightened fears, particularly in places where groundwater is only 3 metres beneath the surface.

The controversy over Melin's involvement and the proposed CAFO divided the community. Some residents called for greater transparency and accountability. Others supported the economic benefits promised by the project. Legal challenges and political disputes arose as the county navigated this contentious issue. In response to the widespread opposition, Burnett County passed a one-year moratorium on large-scale farms in 2019. This delay allowed for further study of the environmental and public health impacts of such operations. Scientists and experts involved in studies supported the fears of locals. One expert confirmed that the smell from the CAFO would be detectable up to 15 and a half miles away. They expected that this would happen at least twice during any given week, depending on how the wind blew.

As of recent reports, with two applications submitted and turned down, the proposed CAFO has not been constructed. The debates over its potential environmental and social effects remain unresolved. The situation has become a case study of how industrial agriculture projects can impact rural communities. Creating fallout for both the environment and politics.

Land Use and Habitat Destruction

The land use associated with pig farming is considerable. It has significant environmental implications, particularly for habitat destruction and biodiversity loss. The expansion of pig farming often involves converting natural habitats into agricultural land. This causes a range of ecological impacts, particularly when forests and wetlands are destroyed.

Most of the land used in pig farming is for growing feed crops, mainly soya and corn, as well as other grains. When farmers convert wild land for this use, the destruction of habitats, critical for wildlife, leads to biodiversity loss and the disruption of ecosystems. Often, this loss of natural habitats is irreversible. There will be the permanent loss of species in an area and the loss of balance in ecosystems. This balance is essential for humans as it maintains processes such as water purification, climate regulation and soil fertility. Insects, with a crucial role within microenvironments, are the first to go. They provide food for small animals and recycle waste, releasing nutrients for flora. Killed during land conversions, the land starts to suffer immediately.

> *Soil structure regulates critical processes such as water, air and solute movement, biological activity and root growth, which directly or indirectly affect the ecosystem services provided by soil, such as crop production, the retention/removal of pollutants, water regulation and carbon sequestration and climate regulation. – Javis, Croucheney et al.* [231]

The intensive nature of pig farming, particularly in industrial operations, degrades soil. Overuse of fertilisers and pesticides in feed production depletes nutrients, reducing soil health. This makes it more difficult to sustain agricultural productivity over the long term. Soil degradation also contributes to other environmental issues. Problems such as erosion, desertification and reduced water retention occur as soil structure changes. Soil can become compacted or over-tilled, causing moisture loss as soil cannot hold rainwater.

A "good" agricultural soil will consist of around 50% solids, 25% air and 25% water, with solids being a mixture of organic material and minerals [232]. This mixture allows for nutrient recycling. Deposits of leaves, dead plants, etc., allow the action of microbial activity and insects, such as earthworms. Plant roots move water and nutrients through soil, while air pockets and moisture stabilise the soil structure. Healthy soil also sequesters carbon and prevents flooding by absorbing water during heavy rain or snowmelt. But agricultural activity prevents plant material from becoming incorporated into the soil. It stops earthworms from being able to recycle nutrients. Heavy machinery compacts soil, squeezing out air, making it hard and impenetrable. These impacts make climate change challenges worse, especially in areas already prone to drought and extreme weather.

Environmental Racism

Environmental racism is the term for systemic discrimination and inequities affecting marginalised groups. It results from policies, practices or decisions that disproportionately expose racial and ethnic communities to environmental hazards. The pollution spans a range of industries, from animal agriculture to petrochemical companies. These affected communities often have less political power. They have fewer legal protections and resources to resist powerful industries. These inequities, entrenched in historical practices, continue to manifest in modern systems. It weighs systems of land ownership, access to resources and policy implementation against them. The effects of the discrimination are far-reaching. They affect economic opportunities, people's right to good health and environmental justice, and it is a worldwide problem.

North Carolina Hog CAFOs and Black Communities

Duplin and Sampson Counties, North Carolina

North Carolina, being the second-largest pork producer in the U.S., has thousands of pig CAFOs. These farms are primarily located in Black, Latino, and Indigenous communities. Residents face pollution and health risks from the industry, with few resources to stop these operations [233].

Air Pollution is released from massive manure lagoons. Ammonia, hydrogen sulphide and methane cause asthma, nausea, headaches and respiratory diseases. **Water is contaminated** with the waste from lagoons. They often overflow during storms, leaking into rivers and groundwater. This leads to **nitrate pollution**, which has been linked to cancer and infant mortality. **Untreated pig waste**, sprayed onto fields, causes bacteria to become airborne. The unbearable smells make it impossible for residents to open windows or enjoy outdoor spaces. Cars, homes and other buildings have to be washed regularly to remove the film of manure misted across towns, carried by the wind.

State and local officials have historically approved these farms in Black and Latino areas. Wealthier white communities will successfully fight against the farms, keeping them far away. Despite community complaints, the state has prioritised industry profits over residents' health.

In 2018, a group of Black residents sued **Smithfield Foods**, the world's largest pork producer, for public health damages. The court **awarded them $473.5 million**. But North Carolina lawmakers, under pressure, later capped the payout. Farm trade groups and politicians, who held a lot of power, openly complained that these suits endangered the industry. They forced a change to North Carolina's right-to-farm law, increasing protection for farmers' lawsuits. The language limited the ability of plaintiffs to seek punitive damages and reduced what qualifies as a nuisance [234].

Iowa's Hog Farms and Indigenous & Latino Communities

Buena Vista, Kossuth and other heavily farmed counties in Iowa

Iowa is the largest pork producer in the U.S. and farms over 23 million pigs in CAFOs. Many of these CAFOs are in areas with large Indigenous and Latino populations. This means that these communities are disproportionately exposed to health risks.

Heavily farmed counties suffer from severe water pollution resulting from manure spraying, contaminating wells and rivers. Nitrate levels in water are high, and toxic algae blooms are common. Workers and residents are victims of **chronic asthma and lung infections** due to constant exposure to **ammonia, endotoxins and dust**. The levels of these illnesses are measurably higher in these communities [235].

Pig farms are the biggest industry in these counties. Many workers are local, but there are also lots of immigrant Latino workers. These people must face, daily, unsafe working conditions, low wages and limited access to healthcare.

CAFOs are often placed in areas where people have fewer resources to fight back. Larger operators, such as Iowa Select Farms, prioritise production over the well-being of these communities. There is also local government involvement. The government creates state policies to protect the pork industry from stricter environmental regulations.

Industrial Pig Farming in China and the Displacement of Ethnic Minorities

Guangxi, Sichuan and Inner Mongolia, China

China produces more pork than any other country. But its large-scale pig farms have forced the displacement of ethnic minority groups, especially in poorer, rural regions [236].

Ethnic groups, including **Uighurs and rural Mongolian herders**, have been **forcibly removed** from their land. This land is then used for large pig farms, with no provision for the people to move to a new area. Water pollution, from pig waste, has poisoned rivers in some areas, causing fish die-off and the loss of clean drinking water. In rural China, this is a vital provision, as there is no municipal water supply. People rely on the freshwater from rivers to drink, cook, clean, fish and grow crops. Respiratory diseases have also become common in farming regions, where people were once healthy.

The Chinese government has decided to place industrial farms in **ethnic minority regions**. In these areas, residents **have no legal power** to resist displacement or pollution. This policy favours wealthier urban areas that enjoy **cheap pork**. The rural communities receive none of the product, but suffer **displacement, environmental degradation** and the **health issues** that result.

The Ethical and Environmental Case for Reducing Pork Consumption

Necessity

There are no places in the world where people **must** eat pigs to survive. In some regions and cultures, pig farming and eating pork are part of their history. Economic and environmental factors also play a role, making pork a key part of their diet. For example, many Pacific Island nations raise pigs for food and ceremonial purposes. In these regions, plant-based alternatives are technically possible, but rarely considered.

Pigs are often used for food as they are efficient at converting feed into meat. They can thrive on agricultural byproducts, food waste, or foraged materials. This has historically made them a practical choice for small-scale farming. In regions with climates or terrains unsuitable for growing diverse crops, people can still raise pigs. These animals don't need large grazing areas, like animals such as cattle or sheep, as they eat most foods. This makes sense for rural or disadvantaged areas, where access to plant-based protein sources like lentils, beans, or soy may be limited.

The eating of pigs, though, is rarely a true necessity. Plant-based agriculture could replace pig farming. This is possible with changes in culture, more access to different foods, and improved global food distribution systems. For example, in some regions of Sub-Saharan Africa, people raise pigs because they adapt well to local conditions. But plant-based diets featuring millet, sorghum and legumes are also common. In the Pacific Islands, while pigs are important in traditional diets, many islanders also rely on taro, breadfruit and coconuts, which could, as part of a diverse diet, sustain populations.

The Environment

For the environment, for the planet, there is a strong case for reducing or eliminating pork consumption. Adopting a vegan or plant-based diet would mitigate the environmental damage from pig farming. At the same time, it would promote the development of sustainable and ethical food systems. It would not be sudden; it would be a process of change. But something similar has already been seen in the expansion of land used to grow crops to produce plant milk. In the United States, almond production grew by 60% from 2010 to 2020. Experts attribute this upsurge to increased demand for almond milk (237).

Plant-based diets are better for the environment than those with pork and other animal products. Switching to plant-based alternatives would help the environment by using less water. It would also reduce water and air pollution, cut down on pesticides and fertilisers, and lessen habitat destruction. Plus, there would be fewer greenhouse gas emissions.

Ethics

Cutting pork consumption has environmental benefits. It also helps with ethical concerns about animal welfare. The conditions that pigs live in on industrial farms are harsh. Animals are subjected to confinement, overcrowding and inhumane treatment. The breeding of pigs, keeping sows confined in crates and pens for cycle after cycle of producing piglets, is cruel. Then there is the method used on boars to harvest their sperm. Outside of pig farming, some might consider this bestiality. But, within the context of farming, this is simply a part of the process. Were this process common knowledge, if the public understood that this was one step of many required to get their bacon or pork chop in the fridge, they might decide to eat something else.

Sustainability

Cutting back on pork would help create more sustainable food systems. A higher demand for plant-based options encourages eco-friendly farming practices. These practices benefit both the environment and our health. they can also drive innovation in the food industry. It could lead to new products and technologies that lessen the environmental impact of food production. They would also support sustainability and food security goals. Companies like Beyond Meat and This are already making strides. They aim to offer plant-based meat alternatives that attract both meat eaters and non-meat consumers.

Health

Farming and consuming pigs pose various health risks to humans. These can be categorised into **occupational hazards** for farmworkers and **health risks** for consumers. The consequences of these can be injury and sometimes death.

Health Risks to Farmers and Farmworkers
Zoonotic Diseases: Pigs can carry diseases that transfer from animal to human to human (zoonoses).

These diseases include:

- **Swine Influenza** (H1N1) causes respiratory illnesses in humans,
- **Brucellosis** is transmitted through direct contact with infected pigs or their fluids, leading to fever, joint pain and fatigue.
- **Leptospirosis** is spread through contact with pig urine, causing liver and kidney damage.
- **Streptococcus suis** is a bacterial infection that can lead to meningitis, hearing loss or even death.

Respiratory Issues: The respiratory tract of both humans and animals is a delicate structure that is easily damaged by the toxic environment within pig farms.

- **Ammonia, hydrogen sulphide and particulate matter** from pig manure and waste lagoons can irritate the respiratory tract. Prolonged exposure leads to chronic bronchitis, asthma, or other lung diseases.

Antibiotic Resistance: Overuse of antibiotics in pig farming contributes to the rise of antibiotic-resistant bacteria (e.g., MRSA). Farmers can contract these bacteria through direct contact or inhalation. Studies show that farmworkers' blood has higher levels of antibiotic-resistant genes than non-farm workers. Their families can also have higher blood levels of these genes than those with no contact with CAFOs or anyone working in them [238] [239]. A 2023 study by World Animal Protection linked **one million human deaths** to antibiotic abuse in factory farming. This is projected to double by 2050.

Injuries and Accidents: Working with pigs poses risks of physical injury. Animals in these situations will bite, kick or crush staff, particularly in CAFOs. In slaughterhouses and meat processing factories, workplace injuries are the norm. A recent study by the U.S. Department of Agriculture (USDA) found high levels of musculoskeletal disorders in the industry. Nearly **42% of pork**

processing workers reported "moderate to severe upper extremity pain" over the previous year [240]. This high incidence is primarily attributed to repetitive motions and high processing speeds. The injuries are so common that staff have become indifferent to their pain. Between 2015 and 2021, **less than 25% of severe injuries** had follow-up inspections by the Occupational Safety and Health Administration (OSHA). This included injuries that resulted in amputations, eye loss, or hospitalisation [241].

Mental Health Impacts: It is also common for workers in large-scale industrial pig operations and slaughterhouses to experience psychological distress. Poor working conditions, exposure to pollution, and ethical concerns about the use of animals and their welfare are all chronic causes.

> *"The first time when I killed it was not easy for me. I feel pity for it. I felt I just wanted to close my eyes, turn around and run away… I feel nothing anymore. In the beginning it was very bad." - Study participant RP9*

In 2016, Victor and Barnard examined slaughterhouse work [242]. *A South African study of slaughterhouse workers* reported severe negative impacts resulting from the work. Slaughterhouses have some of the highest rates of workplace injury. This is due to the nature of the work and the sharp implements that are used. But the study also found that the risk of post-traumatic stress disorder was high. Many workers had regular nightmares that continued to disturb them during the day. Dreams of bodies with no head, being watched by animals. Having animals pleading to be saved or asking, *"Why are you killing me?"* was frequently reported. Deviant and anti-social behaviour of slaughterhouse workers has been documented in many countries. This happens both in and outside of the work setting, becoming a problem for society. Troubling behaviour included substance abuse and domestic violence. The workers reported feelings of shame and fear around their work. They felt that they had to lie to those outside of the industry and not admit to the work that they did. They told of committing domestic abuse, saying, *"I need to hit, especially my girlfriend."* This is not healthy in any context, but in the context of an ingredient on your plate, it becomes abhorrent.

Health Risks from Consuming Pigs

Increased Risk of Chronic Diseases: Processed pork products like bacon, sausage and ham are classified as **Group 1 carcinogens** by the World Health Organization (WHO). They have been directly linked to colorectal cancer. **The International Agency for Research on Cancer (IARC)** have also investigated this. Their findings show that eating 50 grams of processed meat daily increases the risk of colorectal cancer by approximately **18%** [243]. Pig meat is also high in saturated fat and cholesterol. Both of which increase the risk of heart disease and stroke.

Foodborne Illnesses: Raw and undercooked pork needs careful handling due to pathogens such as:

- **Trichinella spiralis** (causing trichinosis) is a parasitic infection that leads to muscle pain, fever, and weakness.
- **Salmonella** and **E. coli** are common causes of food poisoning, causing stomach pain, vomiting and diarrhoea. Illnesses can require hospital treatment and can even lead to death.
- **Hepatitis E virus**, linked to pork liver and other products, the virus is a cause of liver damage.

Toxins and Contaminants: Pork can contain harmful substances. Chemicals like dioxins and environmental pollutants accumulate in animal fat. These chemicals are linked to both cancer and reproductive health issues.

Antibiotic Residues: Antibiotic overuse in pigs can leave residues in meat. When people consume this meat, the antibiotics can cause allergic reactions or disrupt the gut microbiome. This overuse also encourages the growth of antibiotic-resistant bacteria, a global health concern.

Case Study Examples

1. North Carolina Hog Farm Workers – Respiratory and Zoonotic Diseases:
   - Workers in North Carolina's industrial hog farms reported higher rates of respiratory issues. These included chronic bronchitis and asthma, due to prolonged exposure to ammonia and particulate matter. Additionally, outbreaks of **swine influenza** in farmworkers highlighted the risk of zoonotic diseases [244].
2. MRSA Infections from Pig Farms in the Netherlands:
   - Studies in the Netherlands showed that people living near pig farms had a significantly higher risk of contracting **antibiotic-resistant MRSA**. This was linked to the overuse of antibiotics in pig farming [245].
3. Trichinosis Outbreak in the U.S. (1980s):
   - Before modern regulations, trichinosis was more common in the United States. People were becoming ill from eating undercooked pork. A notable outbreak from contaminated sausages in Illinois affected over 100 people. [246].
4. Hog Farm Emissions and Community Health in Iowa:
   - Communities in Iowa living near CAFOs reported higher rates of asthma, headaches and nausea. These were all caused by emissions of hydrogen sulphide and ammonia from pig farms. Lawsuits have highlighted the public health burden posed by large-scale pig farming [247].

Conclusion

Pig farming is a major contributor to environmental degradation. It has significant impacts on greenhouse gas emissions, water and air pollution, land use and biodiversity. The industrial scale of modern pig farming exacerbates these issues. This industry is causing widespread environmental harm. It is also contributing to the global challenges of climate change and resource depletion.

Pig farms house tens of thousands of pigs in a continuous cycle of birth, growth and slaughter. There is no sustainable or environmentally friendly way to manage the waste that results. There is also no sustainable way to produce enough crops to feed pigs farmed on this scale. This is a counterintuitive process in itself. It would be more efficient to feed these crops to people.

Reducing or eliminating pork consumption is crucial to mitigating the damage of pig farming. This shift would protect the environment but also support the development of more sustainable and ethical food systems. Systems that are essential for the health of our planet and future generations.

Choosing veganism over farming and eating pigs is not just a dietary shift. Every year, millions of pigs endure immense suffering, confined to factory farms. They leave the farm to be slaughtered for food that harms the workers, our bodies and the planet. The pollution from pig farms contaminates the air and water, driving climate change and devastating ecosystems. Meanwhile, the demand for pork fuels deforestation, antibiotic resistance and pandemics. Choosing to eat a plant-based diet means a break from a system that exploits animals. A system that protects corporate profits over public health and accelerates environmental collapse.

Chapter 8: Dairy Farming's Dirty Footprint

"Calves are a waste product to us." – A German dairy farm owner

Dairy Farming – An Abbreviated History

The history of humans consuming the milk of other animals spans thousands of years and likely began between 9,000 and 8,000 BCE, in the Fertile Crescent. The Fertile Crescent is a curved region in the Middle East. It was in this area that early humans domesticated sheep and goats, and later cattle. The milk that people drank began as a by-product of animal domestication. Humans milked sheep and goats before cows due to their smaller size and easier handling. Most people could not digest **lactose** (the sugar in milk) during this time. They would lose the ability to produce lactase, the lactose-digesting enzyme, after weaning. But a genetic mutation allowed **lactase persistence**. Dairy farming increased as this mutation spread through populations. This mutation provided a survival advantage, particularly during times of food scarcity. It is still more common in the regions that relied on milk, such as Northern Europe and parts of Africa.

As civilisations advanced, dairy consumption spread and evolved. People developed processes to make fermented dairy products, like cheese and yoghurt. These products had the advantage of reducing the lactose content of milk for those who were lactose intolerant. In Europe, cattle herding expanded around 6,000 BCE. Milk became vital in regions with harsher climates, where agriculture was less reliable. Across Asia, milk became a part of cultures, medicinal practices and religions. The success of cheese-making made the food product a valuable commodity. Being more stable than other dairy products, people would carry cheese to trade.

The industrialisation of dairy farming began during the medieval period in Europe. People who kept cows would produce dairy products to sell. This small-scale production focused on making butter and cheese to preserve milk. By the colonial era, European settlers introduced dairy animals to the Americas, Australia and New Zealand. Historically, milk was absent in these regions. Indigenous diets did not include any dairy due to widespread lactose intolerance. The Industrial Revolution brought technological advancements, like refrigeration and pasteurisation. This made milk into a mass-market product, commodifying cows in the process.

In the 20th century, milk consumption was heavily promoted in Western countries as a vital source of calcium and nutrients. This endorsement was due to the influence and money of the dairy industry. Wanting to increase sales, they pushed the idea that milk was necessary for good health. The development of powdered and condensed milk expanded dairy's accessibility. Public campaigns, such as those promoting milk for strong bones, solidified its place in modern diets. As the dairy market expanded in Western countries, many Indigenous groups and communities in Asia and Africa remained lactose intolerant, consuming little or no milk.

Dairy farming is now often seen as a benign and even necessary part of human diets. However, like other forms of animal agriculture, dairy farming has significant environmental impacts. The production of milk, cheese, butter and other dairy products requires vast amounts of resources. It generates large amounts of pollution, contributing to a range of environmental challenges. Water pollution, greenhouse gas emissions, land degradation and biodiversity loss are all features of dairy farming.

The Process

Dairy farming is built on a cycle of forced breeding, milk production and slaughter. The processes broadly follow those of beef farming.

Dairy cows, as with any mammal, including humans, must give birth to produce milk. This means the use of forced impregnation, usually through **artificial insemination**. Farmers collect semen from bulls and insert it into female cows using specialised tools. This process happens repeatedly throughout a cow's life. She will remain in a state of near-constant pregnancy and lactation.

Once a cow gives birth, after 9 months, her milk begins to flow. But instead of feeding her calf, humans take the milk for their own consumption. Dairy cows are confined indoors or in crowded pastures, depending on the farm. Their lives revolve around repeated pregnancies and milking. This often means standing on hard concrete floors, which can lead to painful conditions like lameness. Repeated milking, up to three times a day, can lead to mastitis (a severe udder infection). Other infectious conditions, such as cowpox, bovine respiratory illness and bovine warts, also occur on dairy farms. These illnesses need to be treated to prevent contagions, including pus, from entering the milk. In veterinary medicine, professionals commonly find blood in milk, particularly after calving [248]. Also, dairy cows are selectively bred to produce unnaturally large volumes of milk. This puts extreme strain on their bodies. Over time, the continuous weight of milk within the udders of cows can lead to distention. Unable to support the udder with the weakened muscles and tendons, they fall downwards. This can result in cows kicking or even treading on their udders as they move around.

After the birth of her calf, a cow goes into a daily routine of milking until she comes into oestrus and is inseminated again. A dairy cow is "spent" by the age of **4 to 6 years**, despite having a natural lifespan of around **20 years** [249]. Years of forced pregnancies and over-milking take a severe toll on her body. She becomes exhausted and prone to infections and lameness. Once her milk production declines, she's no longer profitable. She will go to slaughter, usually becoming low-grade beef, pet food, or leather.

Calves

Since, in dairy farming, a cow's milk is used for human consumption, calves are usually taken away within hours or days of birth. This is an extremely distressing process for a cow; the bond between a cow and her calf is strong. Taking calves away causes cows tremendous anguish. Dairy cows will call for their calves for days following separation. Following this separation, the cow goes back to the herd for daily milking. The fate of her calf depends on its sex.

Male calves are useless in the dairy industry because they don't produce milk. Various regions influence how farmers use male calves, depending on local practices and market demands. But, in many countries, farmers slaughter most males within weeks for veal or kill them at birth as "waste."

In the European Union (EU), farmers raise a significant number of male dairy calves for veal. Approximately six million calves die for veal each year. Top producers are France (more than 1.4 million calves), the Netherlands (1.5 million), and Italy (just under 800,000). Most of the veal produced on the continent is "white veal". This is meat from calves aged eight months or less, fed a low-iron, milk-based diet. This controlled diet keeps the animal's meat tender and as pale as possible. The feed consists primarily of milk replacers, providing protein and energy. The low iron levels induce anaemia, preventing muscle darkening [249]. Other ingredients include: animal fats or vegetable oils (such as palm oil or coconut oil) for calories, vitamins, minerals and lactose from milk or added sugars for energy. Veal calves are usually slaughtered, or "**harvested**", at **16-20 weeks** of age. But "Bob" veal will use younger animals, slaughtering them at 3 weeks old [250].

This processing of veal is a direct consequence of the dairy industry. Male calves are a by-product and are used for meat production rather than milk. Farmers kill those that aren't used for veal very soon after birth, usually by shooting. It is rare to have integration of males calve into beef production systems. Farmers who don't sell calves into the beef or veal system will kill the calves for the sake of the economy. At £9 a head, it is cheaper to shoot a calf than to raise it [251]. Those raised for beef often come from a process of crossbreeding dairy cows with beef sires. This produces calves more suitable for meat production. This can, though, cause problems for the cow. Cross-bred calves may be too large for a natural birth. Caesareans and birthing complications increase cow mortality within these settings.

To prevent the birth of male calves, sexed semen technology allows dairy farmers to breed cows to be females (heifers). This reduces the number of unwanted male dairy calves.

Technicians separate sperm cells carrying female X chromosomes and male Y chromosomes using flow cytometry. This enables breeders to use semen samples that will produce females. The process isn't 100% accurate, with a success rate between **80 and 90%**. This results in stronger, healthier calves more suited for beef production. The calves fetch higher market prices as farmers select sperm to produce offspring with the desired traits [252]. This is also considered to be more environmentally friendly. Reducing unwanted births can lead to fewer calf deaths and lower methane emissions.

Sexed semen, though, is more expensive than conventional semen. Conception rates are less frequent. In the UK, roughly 50% of dairy calf births are conceived using sexed semen. The adoption rate is far lower in less developed countries due to the cost.

In rare circumstances, given the scale of dairy farming, farmers will keep a male calf that shows good genetic potential for breeding. When used on farms, these bulls will impregnate, or service, 50-60 cows in a breeding season. The number of matings is capped; if a bull services too many cows, his sperm count will drop, leading to failures. Bulls used in this way work for one breeding season. Farmers do not like to keep male cows any longer than this, as once a bull reaches two years of age, it becomes harder to handle. For this reason, most farmers use artificial insemination, which is how most male calves are used for breeding.

Female calves, once separated from their mothers, become future dairy cows. Within a day, she moves to an individual pen or hutch. Here she is bottle-fed or given a milk replacer for the first few weeks of life. Many farms use artificial milk, made with milk powder, instead of real cow's milk to maximise profits. This is because, from 1 day old, a calf requires 6-8 litres of milk per day [253].

At around 6 to 8 weeks old, the calf weans from milk, transitioning to solid food, such as calf starter pellets and hay. She may also move to a group housing system with other young heifers. During this stage, her diet will shift to energy-dense grains and forages as she grows. By the time she is 6 months old, she will live in larger pens or pastures, depending on the farming system.

By 12 to 15 months of age, the heifer reaches sexual maturity and is **artificially inseminated (AI)** or bred with a bull. AI is the preferred breeding method in commercial dairy operations. It allows efficiency, genetic selection and reduces the risk of injury from natural mating. Pregnancy in cows lasts about nine months. During this time, the heifer feeds on a nutrient-dense diet to support foetal development. At around 24 months old, she will give birth to her first calf, marking the beginning of her life as a dairy cow.

As of 2024, 18.33 billion kg of dairy is produced in the U.S. annually, with sales of skimmed and low-fat milk totalling 6.9 billion USD

Data from Statista

Immediately after birth, she will have her calf taken away. Just as she was taken from her mother, so that humans could collect her milk for their consumption. She then moves into the milking herd, for milking two or three times a day by machine. To maintain milk production, she is impregnated again within two to three months after giving birth. This cycle of pregnancy, birth and milking continues for several years. When her milk production declines, or if she develops a health issue, such as mastitis or lameness, she becomes unprofitable, or spent and is slaughtered at around 4 to 6 years old.

Dairy farming is often marketed as gentle. But the reality is one of exploitation, suffering and premature death - for both cows and their calves.

Resource Use in Dairy Farming

There is a lot of overlap between the environmental damage of dairy farming and beef farming. This is because both use the process of breeding and keeping cows. Where processes differ in the way that dairy is used.

Water Use

One of the problems with agriculture as a whole, but particularly in animal agriculture, is the transportation of resources and pollution. In the case of pollution, crops grown for feed in one country are then transported to another to feed animals. The second country then produces the manure and associated pollutants. Water used to grow crops travels out of a country in the form of the crop or the feed produced from it. The water, transported to different countries, is embedded within the crop. Even when crops aren't transported, the problem is exacerbated in countries suffering from water scarcity. Both Mexico and India (one of the largest dairy producers in the world) have low water resources. **19%** of the world's water footprint comes from dairy production. **98%** of water use in dairy farming is the result of crop production, and of that, **85%** comes from irrigation alone [254]. Water use in dairy production has three main stages: **on-farm water use**, **feed production** and **processing of dairy products**.

On-Farm Water Use

On-farm water use includes drinking water for cows. This use increases when cows are producing milk, as milk is 87% water. Cows, being large animals, drink large quantities of water. A cow of breeding age will consume between **92-146** litres of water per day [255]. The exact amount depends on the breed of cow, whether they are producing milk and atmospheric conditions, such as hot, dry conditions [256]. Although many dairy farms will house 100-200 cows, dairy CAFOs do exist. These operations can be colossal, housing **3,000 to 10,000 cows** or more. For example, a **large dairy CAFO** could have between **5,000 and 10,000 dairy cows** confined in a barn. They are often in stalls and managed for high-efficiency milk production. Some operations in the **western U.S.** or large dairy-producing regions (like parts of California and Idaho) have very large herds. Taking the median of 119 litres of water per cow, a smaller farm with 150 cows would need 17,850 litres of water daily for drinking. The largest CAFOs with 7,500 cows would use 892,500 litres of water to supply each cow with drinking water. In dry regions needing careful water management, like California, U.S., this can be a considerable burden.

Cattle	Amount of drinking water (litres/day)
Cow with calf	50
Dairy cow in milk	68 - 155
Yearling	24 - 36
2-year-old	36 - 50

The Department of Agriculture, Environment and Rural Affairs [200]

Table 2

Farms use water-based cooling systems to keep milk cool before collection for processing. Technologies are being developed to allow for water recycling. But many of these systems are one-way, with freshwater pumped through and discarded. Hygiene also has high water demands, as milking parlours and equipment need frequent washing. In this part of the process, water flushes manure, cleans barns and disinfects milking systems.

The Water Footprint Network calculated that producing one litre of milk uses about 1,220 litres of water [60][61]. In areas where water is scarce, the high water demands of dairy farming can lead to the depletion of local water sources. This includes rivers, lakes and aquifers. This threatens the availability of water for other agricultural uses. It also endangers the survival of local communities and ecosystems relying on these water supplies.

Dairy Processing

Milk is turned into more than 50 products. These include fresh dairy items and industrial ingredients. Some, like **cheese and butter**, have a much higher water footprint than liquid milk due to the concentration of milk solids. These products are categorised into **liquid, solid, fermented** and **by-product-based dairy** foods. Each of them requires significant amounts of water for processing. The list of products made from milk includes:

- Liquid Dairy Products, such as Whole, Skimmed, Evaporated and Condensed Milk
- Fermented Dairy Products, like Yoghurt, Greek Yoghurt, Kefir and Sour Cream
- Cheese & Curd-Based Products include Fresh, Soft, Semi-Hard and Processed Cheese.
- **Butter & Fat-Based Dairy** products are generally churned milk and include **Butter**, **Ghee** and **Whipped Cream.**
- **Milk Powder & Protein Products** have a wide variety of uses, from baby formula to protein powders. They include **Skim Milk Powder, Whey Protein Powder** and **Casein Powder.**
- **Whey-based products (By-Products of Cheese Making)**, such as **Lactose** and **Whey Protein Concentrate & Isolate**. These are a category of their own due to the number of products made from whey alone.
- Ice Cream & Frozen Dairy like Gelato, Frozen Yoghurt and Dairy-Based Desserts

- Speciality & Industrial Dairy Products include Milk-Based Infant Formula, Lactose-Free Dairy, Casein & Dairy Ingredients and Pharmaceutical & Cosmetic Dairy Ingredients. These products usually need more processing than other dairy-based products.

Water is a part of several steps in producing liquid milk for consumer or commercial use. It is used for feed production, care of animals and hygiene. After milking, water is a part of the pasteurisation, homogenisation and bottling processes of milk. Combined, these processes use **1020 litres of water to produce 1 litre of milk**. Cheese production **concentrates milk solids** while removing water and whey. Because of this, the water footprint of cheese is higher per kg of cheese produced than the same amount of milk. Cheese has a water footprint of **3178 litres to make 1kg of cheese**. This number is then compounded by cheese-making methods, with different cheeses using more water. Harder, aged cheeses, such as cheddar, need more water than softer varieties. Butter has higher water requirements for processing and the multiplier effect of concentrating milk. This results in a water footprint of **5550 litres of water per kg of butter produced** (60) (61).

Production of other dairy products, such as whey, butter and yoghurt, all use water for processing. This includes water use in maintaining hygiene. Little of this water remains in the final product. This is because many of these products result from concentrated milk, and removing water is part of the process.

Dairy plants use **massive amounts of water** to clean equipment, prevent contamination and process waste. Milk is a pathogenic product at the time of milking. **Raw milk can contain various pathogens, contaminants and residues**. These include **bacteria, viruses, fungi and parasites**. Raw milk can also contain manure, skin cells, antibiotics, blood and pus contamination. Milk, being rich in nutrients, is the perfect environment for pathogens to reproduce. Because of this, cleaning of dairy processing plants must be thorough and rigorous. All equipment, surfaces and storage areas must have regular sanitation. This is to prevent bacterial growth, cross-contamination and biofilm formation. Wastewater needs to be treated to **remove dairy residues and contaminants** before release into the environment. But this is not enforced in all dairy farms around the world.

Total Water Footprint of Dairy

When accounting for **feed, farm use and processing**, dairy has an enormous water footprint:

- 1 litre of milk → 1,020 litres of water
- 1 kg of cheese → 3178 litres of water
- 1 kg of butter → 5550 of water

Land Use and Deforestation

Dairy farming requires vast amounts of land for grazing cattle and growing feed crops. In many regions, forests and other natural habitats have been cleared to create space for dairy farms. This leads to significant deforestation and a loss of biodiversity. The expansion of dairy farming into undeveloped areas contributes to habitat destruction and fragmentation. This is having devastating effects on wildlife populations. Furthermore, the loss of forests reduces the planet's capacity to sequester carbon, worsening the impacts of climate change.

Changes in land use, such as the conversion of forestland to pasture or grassland to cropland, release greenhouse gases into the atmosphere. When people cut down trees or clear land of wild vegetation, the organic matter left rots. This process happens both above and beneath the ground in a process of oxidation. This releases gases, including carbon dioxide and nitrous oxide, into the air. Experts estimate that deforestation in Brazil releases 37,000 kg

CO_2eq. per hectare. Deforestation in Argentina releases 17,000 kg CO_2eq. per hectare. The clearing of shrubland in Argentina releases 2,200 kg CO_2eq. per ha. Where people farm cattle, what cows eat and where that food comes from, has a huge impact on the damage that results. In dairy farming, all this damage is embedded in the products produced. For example, a dairy cow raised in the UK that grazes on grassland with some soya feed has fewer water inputs and less land use than a cow raised in the USA. In the United States, cattle remain on feedlots and feed on a diet of soya grown on land cleared from a rainforest. This does not make the UK method environmentally friendly, or even neutral. The inefficient process of farming animals to make products from their bodies is damaging, regardless of the process. In the above scenarios, another confounding factor is that grass-fed cows release 65.4% more CO_2eq/kg than those fed on soya and grains [257].

"intensive systems often require more inputs to manage waste. Degradation of ground and surface water from waste contamination is a potential social and economic cost of intensive systems. Unless governments can enforce regulations to monitor and to control nutrient losses in animal waste, producers may choose the lower cost option of no control, undermining yet another policy objective to decrease environmental degradation from livestock production." - Nicholson, Charles F. et al

Nicholson, Charles F. et al. wrote a paper, Livestock, Deforestation, and Policy Making: Intensification of Cattle Production Systems in Central America Revisited, in 1994. In it, they argued that the answer to deforestation for farming cows was to intensify farming processes. *Some analysts have suggested that substantial intensification of beef and dairy systems could curb deforestation, by reducing the additional lands needed to produce greater supplies of these commodities"*. However, as the investigation examined the processes of dairy production, patterns emerged. They saw how people migrated to forested areas for work. That government policies controlled markets, land use and regulations. Through their work, it became clear that packing more animals onto a piece of land and feeding them high-calorie diets wasn't a simple fix [258].

Even outside of the deforestation in rainforest regions, the land used for dairy farming is often land that had been wild. Clearing of land, whether grassland, wetlands or bushlands, releases carbon into the atmosphere. It introduces pollution to the land, air and water and impacts the quality of the lives of residents. The conversion of wild or semi-natural areas to dairy farming has led to significant environmental and social challenges. Regions like New Zealand, Ireland, Australia and parts of the United States are examples of this.

New Zealand: In New Zealand, farmers have cleared, or converted, vast areas of native tussock grasslands and indigenous bush into dairy pastures. Native flora and fauna, such as wildflowers, have declined as dairy farming replaced complex ecosystems with monocrop pastures. Ryegrass, not native to New Zealand, is now planted for cows to graze on. Ryegrass is an invasive species. It outcompetes native plants, dominating whole areas. Bird populations have declined. Species of rail and heron have migrated away from cleared lands that no longer support their natural behaviours. Intensive dairy farming has caused E. Coli contamination of waterways. Increased nutrient runoff (nitrogen and phosphorus) in rivers and streams poisons the water. The resulting algal blooms and degraded aquatic habitats have harmed aquatic plants and animals [259].

Local communities and indigenous groups have raised concerns over these problems. Water quality and allocation have become complicated as dairy farming competes with traditional water uses. Also, while dairy farming has boosted the economy, it has led to changes in rural community structures. The industry has created a reliance on a single industry that is vulnerable to market fluctuations. If dairy farms were to close or move, the economy of the community would be massively reduced. This has the potential to cause poverty in some groups where dairy jobs are the main or only source of income.

As of 2024, The value of dairy production in the UK is worth £6 billion, with 14.9 billion litres being produced annually

Data from Statista

Ireland: In Ireland, traditional mixed farming systems and native hedgerow landscapes have given way to dairy-intensive farming. Regions like the Midlands and parts of the West have seen the greatest increase in dairy farming. The shift to large open dairy pastures has reduced habitat complexity and reduced wildlife diversity. Many miles of woodland and hedgerows, which were corridors and habitats for many animals, are now gone. This has caused the number of farmland birds that nest in hedgerows (e.g., skylarks, linnets) to decline. Small mammals, such as voles and shrews, have moved. A range of invertebrates that thrived in varied microhabitats have seen their numbers reduced. Elevated levels of nitrates and phosphates from fertiliser and manure have caused water quality issues. These are affecting local waterways and coastal ecosystems [260].

Changes in land use have transformed rural landscapes and livelihoods. These changes have created tensions between traditional farming practices and modern, intensive agriculture. Increased dairy production has strained local infrastructure. Roads and waste treatment systems have an increased workload, impacting residents' quality of life. People take offence at large milk tankers on previously quiet roads.

Australia: In parts of Australia, areas that once supported native bushland and wetlands have been cleared for dairy pasture. Regions like Victoria and New South Wales have lost vast areas of wild land. This clearing of native vegetation has disrupted local ecosystems. It has caused great losses of biodiversity and altered soil structure and function. Native species, such as the Eucalyptus, are replaced with monocrops of grasses. Native bird species that depended on diverse bushland and wetland habitats have declined. As feeding and nesting grounds disappeared, birds like honeyeaters, parrots and waterbirds left. The populations of small marsupials, native rodents and reptiles have shrunk. Wetland conversion and changes to natural water flow have harmed frogs and other amphibians, who need specific breeding environments. A reduction in native vegetation has disturbed a host of insects. This, in turn, affects pollination and the broader food web [261].

Dairy farming in these drier regions often relies on irrigated systems. This can lead to over-extraction of water and increased salinity or contamination of water bodies. As water becomes scarcer and polluted, communities must balance the needs of agriculture with domestic and environmental water uses.

Pollution from Dairy Farming

The pollution resulting from the farming of cattle is universal. This is whether farmers raise animals for meat or dairy. Either way, dairy cows are slaughtered for meat when they are spent.

Air pollution largely results from manure and enteric fermentation in the guts of cows. Dairy products account for 4% of global man-made greenhouse gas emissions. Of that, **93%** production is on the farm [257]. Land becomes polluted with manure from animals defecating as they graze or from applications of manure to fields. Water pollution comes from several sources. This is the runoff of nutrients and waste from cows, farms and processing plants.

In 2007, the dairy sector emitted 1969 million tonnes CO_2-eq [±26 per cent] of which 1328 million tonnes are attributed to milk, 151 million tonnes to meat from culled animals and 490 million tonnes to meat from fattened calves. - Food and Agriculture Organization of the United Nations

Air Pollution and Ammonia Emissions: Dairy farming contributes to air pollution through the release of ammonia from manure. Ammonia, combined with other pollutants in the atmosphere, forms fine particulate matter. These particles harm human health when inhaled. High concentrations of ammonia also lead to the acidification of soils and water bodies. Lowering the pH of water and soil damages ecosystems and reduces agricultural productivity.

The odours associated with dairy farms, particularly those in intensive farming operations, can be pungent. These smells affect communities, lowering life quality. Conflicts over land use and expansions to farming operations also make life difficult, and it is not unusual for locals to make noise complaints during the breeding season. When farmers separate cows from their calves, cows cry out. These cows are calling for and looking for their calf. This can go on for several days until cows give up and stop. The volume of a cow's call can be surprisingly loud; they vocalise in a range between **85 and 90 decibels (dB)**. To add context, sounds around 85 - 90 dB are comparable to the noise of heavy city traffic or a lawn mower at close range. When a herd of cows call out, especially when people are trying to sleep, it can disturb the peace. This noise can become a form of noise pollution.

Dairy farming is a significant source of greenhouse gases, particularly **methane (CH_4)** and **nitrous oxide (N_2O)**. Cows produce methane during digestion, and it also results from manure management practices. Methane is a potent greenhouse gas, having a global warming potential many times greater than carbon dioxide.

Nitrous oxide results from the application of manure and synthetic fertilisers to fields. Feed crop production exacerbates this problem due to its large scale. Like methane, nitrous oxide is a powerful greenhouse gas.

Soil Erosion and Sediment Pollution: Manure management in dairy farming usually involves the spreading of manure over fields. If cows remain inside, farmers collect and spread the manure. If cows have more time in an open pasture, nature manages the manure. But this can be a mammoth task as the production of 1 litre of milk results in 8.5 litres of slurry, liquid manure [262]. When cattle graze, their manure deposits directly onto the land as they roam. The greenhouse gases released as the manure decomposes, including nitrous oxide, can negate any ecological benefits, such as fertilising soil.

Overgrazing is also a problem. Cows go through routine rotations to different areas of land to avoid this. Overgrazing strips vegetation and exposes soil to erosion. This can happen when the concentration of cows is too high. Leaving cows in one area for an extended period can also result in overgrazed land. This process can allow wind and water to carry away the topsoil, the most fertile layer. The reduction in topsoil harms the land's ability to support future crops or pastures. Soil can also become compacted through the trampling of cows. This removes air pockets from the ground and reduces the soil's ability to store water. Compacted soils are also hard for plant roots to penetrate. Overgrazed and compacted land can remain barren long after cows have moved on. Over time, these processes can lead to **desertification**. This is an acute problem in semi-arid regions. In these areas, grazing cattle can quickly push ecosystems beyond their recovery threshold.

Water Pollution and Nutrient Runoff: Dairy farms produce large quantities of manure. This waste must be properly managed to avoid water pollution. Applying manure to fields as fertiliser in excessive amounts, or during heavy rains, causes nutrients like nitrogen and phosphorus to leach into nearby water bodies. This nutrient runoff can cause eutrophication, leading to algal blooms, dead zones and the loss of aquatic life.

Dairy farms can also contaminate water sources with the pathogens, hormones and antibiotics in manure. This contamination poses risks to both human health and the environment. It can lead to the spread of disease and the development of antibiotic-resistant bacteria.

Whey

Several phenomena are particularly characteristic - or even unique - to dairy farming. These come from the nature of continuous milk production, the specific management practices required for lactating animals and processing milk into dairy products. An example of this is whey production. Cheese production uses approximately 35% of milk globally [263]. Of that, 80 - 90% of the milk becomes whey, generating an estimated 180 - 190 million tonnes of waste whey globally each year [264]. Whey is the liquid byproduct of cheese-making and dairy processing. It is a nutrient-rich waste stream that is, in part, further processed to produce a list of products. But what remains is a waste product that can pose significant environmental challenges if it isn't treated. Without a sustainable way to handle whey, it becomes a major pollutant in the dairy industry. A lot of whey, in liquid form, is thrown away as wastewater. This leads to serious harm to the environment.

Whey contains roughly 55% of the nutrients found in milk. This means it has a high concentration of organic compounds such as lactose, proteins, fats, vitamins and minerals [265]. This results in high levels of **Biochemical Oxygen Demand (BOD)**, often in the range of 30 - 80 g O_2/L. It also has a high **Chemical Oxygen Demand (COD)**, reflecting the amount of oxygen required to oxidise these compounds. The high nutrient content, particularly carbohydrates and proteins, makes whey an excellent growth medium for microorganisms. This is particularly problematic when discharged untreated. Whey usually has an acidic to neutral pH. But this can vary, depending on the cheese-making process. These variations complicate the requirements to make waste safe for discharge.

When milk processors release whey into rivers, lakes, or soil without treatment, the high BOD and COD cause rapid microbial growth. This process consumes dissolved oxygen, leading to hypoxic conditions (low oxygen) in aquatic ecosystems. What follows causes fish kills and a loss of aquatic biodiversity. The nutrient load - especially lactose and nitrogenous compounds - stimulate excessive algal growth. Subsequent algal blooms and die-offs further exacerbate oxygen depletion and lead to harmful eutrophication.

If farmers apply whey as a fertiliser without proper management, the high nutrient content can alter soil chemistry and microbial communities. If dairy processors dispose of whey on land without care, the high nutrient content can damage soil health. Overapplication may lead to nutrients leaching into groundwater, contaminating drinking water supplies. Also, when applied to land, the microbial breakdown of whey produces malodorous compounds. In other words, it stinks. Nuisance odours can hover in the vicinity of dairy processing plants, making the whole area unpleasant.

Vulnerable to the Future

Dairy farming is linked to climate change, both as a contributor to and a victim of its effects. The industry's greenhouse gas emissions speed up global warming. The warming climate then poses challenges to the sustainability of dairy production. As mentioned earlier, dairy farming is a significant source of greenhouse gases, methane and nitrous oxide. The dairy industry's large-scale production of these gases contributes to the effects and acceleration of global warming. The carbon footprint of the entire process of dairy products is also huge. It starts with feed crop cultivation, processing and transport. There is the breeding and maintaining of cows. Milk needs collection, processing and transportation with a reliable cold chain at every stage.

Dairy farming is highly vulnerable to the effects of climate change. Regions where water resources are already stressed are under great strain. Changes in temperature and rain patterns can affect the availability of water for both cattle and feed crops. When farmers must cut back on feed or feed quality, milk yields drop. This reduced availability also pushes the prices of feed up. Higher prices, combined with lower income from reduced milk yields, cause financial hardships for farmers. Climate change can also exacerbate the spread of diseases and pests. Warmer temperatures make ideal breeding conditions for bacteria, pests and parasites. In these conditions, diseases affecting both cattle and crops can be hard to control. Compounding this, in regions that suffer from heatwaves, heat stress impacts animals' health. Heat stress can reduce milk yields. It leads to increased water consumption on farms for drinking. In severe cases, cattle will need veterinary care. The effect of this is one of reinforcement. Dairy farming contributes to climate change. The repercussions create the need for increased resource inputs so that dairy farming can continue. This strains already diminished resources while poisoning what is left.

Government Cheese

Government cheese is a phenomenon of the United States government. It was originally distributed by the government as part of food assistance programs. But it quickly became a shorthand way of saying that you grew up poor.

"Had second-hand clothes, and we rode thrift store bikes.

I ate government cheese, and it was all about the lifestyle, it was right."

"C.R.E.A.M." by Wu-Tang Clan

"I been a lot of places, but I ain't never been to the White House

I been on the grind, government cheese, I don't kind."

"Fire Squad" by J. Cole

Government cheese is a processed cheese product that became a symbol of federal aid during the late 20th century, particularly in the 1980s. It originated in a combination of excess dairy production, policies that supported American farmers and the need to address food insecurity.

Government cheese started in the 1970s. This is when the U.S. government helped dairy farmers by subsidising them to keep milk prices steady. These subsidies led to the overproduction of milk. To prevent waste, dairy producers processed the milk into cheese, butter and powdered milk. This led to a surplus of these more shelf-stable products. To keep the dairy market operational, the government started buying the milk products. By the early 1980s, the government had purchased billions of pounds of dairy products. They stored these vast amounts of dairy products in warehouses across the country. The government has no use for the cheese and other items. Storing them in temperature-controlled warehouses was costing money. But it was still food and could not be wasted. This is when the federal government decided to distribute the excess cheese as part of food aid programs.

President Ronald Reagan's administration formalised the distribution of government cheese in 1981. They added the cheese, a processed blend of cheddar and other varieties, to commodity food programs. This included the Temporary Emergency Food Assistance Program (TEFAP). It was also distributed to low-income households, food banks and school lunch programs. The cheese provided an affordable source of protein and calories. But it was often criticised for its overly salty and rubbery texture. Even so, it became a staple in many American homes and a cultural touchstone for communities that relied on government assistance.

Government cheese distribution peaked during the **1980s and early 1990s**. But surpluses declined with better management of dairy production and subsidies. Additionally, changes in food assistance programs allowed recipients to buy their own food. This reduced the need for direct commodity distribution by the government. By the late **1990s**, the program had phased out, and the iconic blocks of government cheese became a thing of the past.

Although government cheese is no longer distributed, its legacy endures. It is a symbol of government intervention in agriculture and food security. It is also a symbol of the economic struggles faced by many Americans during its heyday. Today, it remains a nostalgic reference in popular culture, often evoking both fond and bittersweet memories for those who relied on it.

AS OF 2024, 685.91 MILLION KG OF DAIRY IS PRODUCED IN SWEDEN ANNUALLY, WITH EACH PERSON DRINKING, ON AVERAGE, 88.5 LITRES OF MILK EACH YEAR

DATA FROM STATISTA

Schools, Hospitals and Prisons

Governments distributing milk and dairy products in schools, hospitals and prisons is a multifaceted idea. It attempts, through buying and selling dairy, to solve several problems at once. It aims to meet nutritional needs, support agricultural policies (including surplus management) and provide affordable food in institutions. The practice also serves economic and political functions, such as ensuring stable prices for dairy farmers.

In the UK, the government's School Food Standards recommend including a portion of dairy every day. They consider this ideal as part of a balanced diet for children [266]. Many people believe that dairy products like milk are vital to good health. This is because they contain essential nutrients such as calcium, proteins, vitamin D and potassium. This makes sense as cows make milk to turn a calf into a cow within the space of a year.

Schools, hospitals, prisons and other government institutions often serve populations that don't have access to healthy or balanced diets. Providing dairy ensures that these groups receive some of these essential nutrients.

In many countries, dairy farming is heavily subsidised by the government. In these countries, dairy farming wouldn't be profitable without the payments. Without these payments, farmers would have to increase their prices to continue trading. Government subsidies reduce dairy product costs. This makes them affordable for individuals and public institutions. In institutions like prisons, budgets for food are often tight. Dairy products, being inexpensive, can provide a large amount of calories and protein at a low cost.

One significant factor in the distribution of milk and dairy products in public institutions is the need to manage surpluses. It is common for dairy farmers to produce more milk than they can sell in the open market. This is due to government actions that encourage overproduction, such as subsidy payments. The government then buys the surplus and distributes it between food assistance programs, schools, hospitals and prisons. These government actions are, in large part, driven by the dairy industry itself. Each country has its own organisation that acts on behalf of the dairy industry. The United States has *the National Milk Producers Federation (NMPF)*. In Europe, the *European Dairy Association* represent the industry. Australia has the *Australian Dairy Farmers*. These organisations ensure that the dairy industry is not harmed by government policies. They make financial donations to political parties and campaigns of individual politicians. These donations work to garner favour in politics. They then use this favour with lobbying power to influence policies and dairy-related legislation.

By integrating dairy into public institutions, governments provide a secure market for dairy farmers. This guarantees that their production continues to be financially viable. Governments may also use these public institutions to promote the consumption of dairy products. This reinforces the idea that milk is a vital part of a balanced diet. Public health campaigns have historically emphasised the importance of milk for bone health. This idea has led to the widespread acceptance of dairy as an essential food. This all leads to greater milk production. More cows, more artificial insemination, more unwanted calves, more deforestation, more pollution.

Milk in UK Schools

The UK government began providing milk in schools in the **early 20th century** to address malnutrition in children. During this time, poverty was widespread. Many children suffered from undernourishment and health problems. These included rickets, a bone condition caused by vitamin D deficiency. Many believed milk was an ideal supplement as it is nutrient-dense, providing calcium, protein and vitamins.

In **1906**, the Liberal government introduced the **Provision of Meals Act**. This allowed local authorities to provide meals for children. Building on this, in **1946**, the government passed the **School Milk Act**. This guaranteed free milk for all schoolchildren under the age of 18. Each child received a daily one-third-pint bottle of milk. This initiative became a cornerstone of post-war social policies that improved public health and welfare.

The program continued until the **1970s**. In **1971**, Prime Minister Edward Heath's government stopped free milk for children over seven years old. This sparked the phrase "Margaret Thatcher, Milk Snatcher," as Thatcher, then Education Secretary, oversaw the policy change. The milk provision was further reduced over time due to budget constraints and shifting priorities. By the **1980s**, policies limited free milk to younger children. Today, schools encourage parents to pay into a system that provides a daily carton of milk to primary school-aged children (ages 5-11).

None of the policies made any allowance for children who were lactose intolerant or allergic to milk. Pressure has since forced some schools to provide, or allow parents to provide, dairy-free alternatives for their children. But this can still be a fight due to the ingrained belief that dairy is best.

Milk and Cheese in U.S. Schools

In the United States, milk and cheese also became staples in school meal programs. This was due to government policies supporting the dairy industry, combined with efforts to combat childhood malnutrition. During the **Great Depression (1930s)**, the federal government purchased large quantities of surplus agricultural products, including milk. This was to stabilise food prices and support struggling farmers. Schools and other institutions received the surplus products, marking the beginning of government involvement in school nutrition.

The U.S. government established the modern **National School Lunch Program (NSLP)** in **1946**. Malnutrition among military recruits during World War II was the motivation for the project. The program added milk as a key component of school meals due to nutrient content. Similarly, the **Special Milk Program**, introduced in **1954**, had the specific aim of increasing milk consumption in schoolchildren. They achieved this by subsidising its cost, making it cheaper for parents. These programs were often accompanied by educational campaigns. They promoted milk as essential for strong bones and overall health. Most of the funding for the campaigns came from dairy industry organisations.

Cheese also became prominent in school meals. This peaked during the **1980s**, when the U.S. government faced a dairy surplus crisis. In support of the dairy industry, the government supplied surplus cheese ("government cheese") to schools and low-income families. Food suppliers incorporated processed cheese products into school lunches. Cheese appeared in items like pizza, sandwiches and casseroles. They sometimes forced cheese into foods that would have been cheese-free.

In both the UK and the USA, the emphasis on milk in schools has faced criticism over time. Some experts have questioned the necessity of dairy for all children. This is in part due to some people not being able to tolerate dairy. A large section of the American population is lactose intolerant. This is particularly prominent in African, Asian and Indigenous groups. Concerns about the environmental impact of dairy farming have also influenced debates about the role of milk in schools. The growing availability of plant-based alternatives strengthens the arguments against milk.

In recent years, both countries have seen shifts toward more inclusive and sustainable school meal policies. For example, some U.S. schools now offer plant-based milk alternatives. In the UK, there is a growing push to include non-dairy options in milk programs and on lunch menus. Despite these changes, milk remains ingrained in school nutrition policies. This reflects how ingrained milk is in culture and how powerful lobbying groups are.

Dairy in Prisons

In prisons, dairy products such as milk, cheese and butter are often included in meals. This is due to their affordability, long shelf life (especially processed dairy) and nutrient density.

U.S. Prisons: Milk and dairy are commonly served in U.S. correctional facilities. Federal regulations require prison meals to meet minimum nutritional standards. They must also include the nutrients vitamin D and calcium. Prisoners are rarely able to spend enough time outside to receive adequate sunlight to make this vitamin, so they add dairy to provide this. In some prisons, meals include mandatory servings of milk or processed cheese. Milk is given in cartons as a drink. It is usually given for breakfast or as part of casseroles and sandwiches. While inmates may refuse food, alternative options (e.g., non-dairy milk) are often unavailable, with them only being allowed to make requests for religious or medical reasons. For example, lactose-intolerant inmates may need to prove their condition to receive alternatives. Vegan or plant-based meals are becoming more available in some states. But this is largely due to lawsuits and advocacy efforts rather than having suitable alternatives readily available.

UK Prisons: Like the U.S., dairy is a regular component of prison meals in the UK. However, inmates can have more flexibility to opt for non-dairy options. Reasons considered include medical conditions, ethical reasons (e.g., veganism), or religious practices (e.g., certain dietary restrictions in Islam or Judaism). Vegan meal options are also easier to find in UK prisons as part of broader efforts to accommodate dietary diversity.

Dairy in Hospitals

Dairy has been a longstanding part of hospital menus, often included in patients' meals as a source of calcium, protein and calories. However, policies around hospital diets vary widely by country, hospital system and patient needs. Milk is often served with meals. Dairy products, like yoghurt and cheese, are part of "soft" or "nutrient-rich" meal plans for patients recovering from surgery, severe illness, or malnourishment. These foods are seen as easy to digest and palatable for many patients.

In modern hospitals, dietary accommodations are typically available for patients who are lactose-intolerant, allergic to dairy, or following a vegan or plant-based diet. But each hospital will set its own policy. Some hospitals will only offer plant-based options like soy or almond milk if they are asked for. Also, not all hospitals have a variety of alternatives.

Advocacy for healthier, plant-based hospital meals is gaining traction in many countries. In the U.S., organisations like the *Physicians*

Committee for Responsible Medicine (PCRM) have pushed for reduced reliance on animal products in hospitals. Citing links between dairy consumption and chronic diseases, like cardiovascular issues and certain cancers, they are asking for policy changes. There are also some hospitals that now offer plant-based meals as a standard option while reducing the amount of dairy in their menus.

"Globally, the dairy sector is probably one of the most distorted agricultural sectors: producer subsidies are in place in many developed countries, encouraging surplus production, export subsidies are paid by governments to place the excess production on the world markets," - Knipps 2005 [267]

Future Directions

There is growing momentum to reduce dairy consumption in institutional settings. This has been driven by concerns about health, sustainability and inclusivity. Increasingly, prisons and hospitals are incorporating plant-based options into their menus. As awareness of lactose intolerance, veganism, and the environmental impact of dairy farming continues to grow, institutional reliance on dairy will likely diminish. The pace of this change will depend on policy decisions, budgetary constraints and advocacy efforts.

Got Milk

The *Got Milk?* Campaign was one of the most famous and long-running advertising campaigns in the United States. Launched in 1993 by the California Milk Processor Board (CMPB), the campaign was later adopted nationwide by the Milk Processor Education Program (MilkPEP). It featured celebrities and athletes with milk moustaches, along with the famous tagline "*Got Milk?*"

By the early 1990s, the dairy industry had to act as milk consumption in the U.S. was declining. A rise in the popularity of soda and other drinks outcompeted dairy for sales. The plan of action was a massive marketing campaign; it was going to make milk seem essential to everyday life. The campaign was also going to emphasise its necessity for strong bones and good health, and focused on two key strategies:

1. **Celebrity Endorsements** – The ads featured A-list celebrities, athletes and pop culture figures sporting milk moustaches to associate milk with success and desirability.
2. **Humour and Deprivation Strategy** – Advertising companies wrote funny ads. They depicted people suffering the consequences of not consuming milk. One had a man unable to say "Aaron Burr" because his mouth was full of peanut butter without milk to wash it down. In another a man's arms snapped off as he tried to move his wheelbarrow; he didn't drink milk.

The campaign was a massive success in terms of branding. By the late 1990s and early 2000s, *Got Milk?* had become a household phrase. But despite its popularity, the campaign had several issues and criticisms:

1. **Health Misinformation** - The ads aggressively promoted the idea that milk was necessary for bone health and wellness. Modern research shows that consuming a lot of dairy does not always lower the risk of osteoporosis. Some findings suggest

dairy may even increase the risk of some health issues, such as certain cancers.
2. **Lactose Intolerance and Racial Bias** - The campaign ignored the fact that a large percentage of non-white populations suffer from lactose malabsorption or are lactose intolerant. People of African, Asian and Indigenous descent most often cannot digest the sugar. Critics argued that promoting milk as a universal health food was racially insensitive and misleading.
3. **Dairy Industry Ethics** - The campaign masked the cruelty of the dairy industry. There was no mention of the processes involved. The forced impregnation of cows, the separation of calves from their mothers and the eventual slaughter of "spent" young dairy cows. As awareness of these issues grew, animal rights activists heavily criticised *Got Milk?*
4. **Milk and Acne/Health Risks** - Studies have linked dairy consumption to increased acne and other health risks. As plant-based options became popular, people started to question whether dairy was really as good as claimed.
5. **Declining Relevance** - As more consumers switched to plant-based milk, the campaign lost effectiveness. By 2014, *Got Milk?* was officially retired in favour of a new campaign, "*Milk Life,*" which focused more on protein content.

The *"Milk Life"* campaign, created by MilkPEP, shifted the focus from celebrity endorsements to milk's nutritional content. The push this time was the protein content. By the early 2010s, dairy consumption was still losing popularity. At the same time, plant-based milk alternatives were gaining popularity. The *Got Milk?* Campaign, despite its cultural impact, wasn't reversing the trend.

The campaign emphasised that milk contains 8 grams of protein per serving. It was portrayed as a natural energy booster for an active lifestyle. The ads featured everyday people engaged in physical activities. A splash of milk was artistically edited onto their bodies to symbolise movement and strength.

The campaign was not a success. It completely failed to capture the cultural moment in the way *Got Milk?* did and dairy sales continued to decline. Consumers were becoming more sceptical of milk's health claims. Plant-based alternatives kept expanding in availability and variety, giving consumers options. As these alternatives became available in supermarkets and even coffee shops, people found that they felt better without dairy.

Although *Milk Life* is still around in some form, it never reached the iconic status of *Got Milk?* This forced the dairy industry to shift strategies again to their perceived enemy. This time, they attacked plant-based alternatives. One strategy was lobbying against terms like "almond milk" and "oat milk."

Dairy is Healthy?

Milk and milk products are marketed as healthy and even necessary. Dairy industry organisations make sure of that. Doctors and the government tell people that milk is a source of vitamin D and calcium. That they should consume some form of dairy every day. Even the British Heart Foundation advocates for including dairy: "Eight per cent of our food should be made up of dairy products and alternatives". It is important to note the use of "and" alternatives, not "or" alternatives, the implication being that dairy is vital.

> *"What are the unique nutrients that dairy has that nothing else has? Nothing,"* - Christopher Gardner, *professor and nutrition researcher, the Stanford Prevention Research Center.*

The amount of vitamin D and calcium in milk varies by product. The type of milk, the producer and whether the product is fortified all impact values. For example, Cravendale Whole Milk contains 1 µg of vitamin D per 100 ml and 124 mg of calcium. Arla B.O.B Skimmed Milk contains 147mg of calcium with no vitamin D, but Arla Big Milk has 122 mg of calcium and 2.2µg of vitamin D.

Saturated Fat

Milk contains significant amounts of **saturated fat** and **cholesterol**. Both of which are linked to an increased risk of cardiovascular disease. Whole milk contains around **3.8 grams of saturated fat per 200 ml**. This is nearly a quarter of the recommended daily limit for saturated fat based on a 2,000-calorie diet. Saturated fat has been shown to raise **low-density lipoprotein (LDL) cholesterol**, often referred to as "bad" cholesterol. LDL cholesterol contributes to plaque build-up in arteries and increases the risk of heart attacks and strokes. Some argue that the impact of saturated fat on heart disease is complex and depends on overall eating patterns. But leading health organisations, such as the **American Heart Association,** still recommend limiting intake.

Milk contains **dietary cholesterol**, with whole milk providing around **14 milligrams per 100ml** [269]. Dietary cholesterol's impact on heart disease has been debated. However, research shows that a high intake of cholesterol and saturated fat can worsen lipid profiles. This increases cardiovascular risk, especially for those with existing metabolic problems.

Cancers

There is a growing body of work showing an association between dairy consumption and some cancers. The link is not clear. Some investigations show a definite correlation between cancer rates and increased consumption of dairy. Some show no link at all, while others find a negative correlation, that increased dairy leads to reduced cancer risk. There are some cancers, including breast and prostate cancer, that have a more established association. A study by Kakkoura et al. (2022) examined this in a cohort from China of more than 500,000 people [268]. They found that

"higher dairy consumption was associated with a higher risk of overall cancer, liver cancer, female breast cancer..., with each 50g/day higher usual intake being associated with 7%, 12%, 17%... higher risks, respectively".

China has lower rates of dairy consumption compared to Western countries, where most studies of this type are conducted. This made the findings of greater interest in medical publications. This is because the distinction between those who consume dairy and those who don't was more pronounced. Out of this cohort, taken from a variety of regions and economic backgrounds, 20% of the participants consumed dairy products regularly (primarily milk). 11% consumed dairy products monthly, and 69% were non-consumers. The average consumption was 38g per day in the whole study population and 81g per day among regular dairy consumers. Participants from the UK Biobank study had an average consumption of around 300g per day.

During the study period, they recorded 29,277 new cancer cases. The highest rate was lung cancer (6,282 cases). This was followed by female breast (2,582 cases), stomach (3,577 cases), colorectal (3,350 cases) and liver cancer (3,191 cases). They found no association between dairy intake and prostate cancer or any other type of cancer they investigated.

Lactose

Some people have no tolerance for milk. Immediately after consuming dairy, they experience symptoms like bloating and stomach pain, followed later by gas and diarrhoea. These symptoms are the result of **lactose malabsorption**. Lactose is the sugar found in milk and is present in the milk of most mammals, including humans. Lactose malabsorption is a condition where the body has difficulty digesting lactose. **Lactase** is the enzyme that breaks down lactose into its absorbable components: **glucose** and **galactose**. Those with low levels of lactase cannot do this. The ability to produce lactase is usually lost after weaning, when infants stop drinking milk from their mothers. When lactose isn't digested, it moves into the colon, where gut bacteria ferment it. If this happens, but the person has no symptoms, they have lactose malabsorption. If it leads to symptoms such as pain and gas, it is lactose intolerance. The number of people with lactose malabsorption is higher than that of those with lactose intolerance.

Globally, the percentage of people who are lactose intolerant is ≈65%, but this number can be very different between countries. For example, in Denmark and Ireland, the percentage of people affected is 4%. In South Korea, Yemen and Ghana, 100% of the population has no tolerance for milk.

Skin Disorders

There is growing evidence linking dairy consumption to various **skin disorders**, particularly **acne, eczema and rosacea**. While scientists continue to study the exact mechanisms, researchers suggest that hormones, growth factors and inflammatory responses may play a significant role.

Numerous studies, such as one carried out by Ismail et al. (2012) with young Malaysian people, have found an association between dairy intake and **acne breakouts**. The most affected group is adolescents and young adults [269]. There are **hormones and growth factors** in milk. These include **insulin-like growth factor 1 (IGF-1)** and **bovine hormones**. It is believed that these hormones stimulate **sebum (oil) production**, increase **inflammation** and trigger **skin cell overgrowth**, contributing to acne. **A 2018 review of 78,529 individuals** found that **any dairy consumption** increased acne risk by **25%**. Drinking **low-fat or skimmed milk** had a stronger association with breakouts. Following this, a **2020 systematic review** confirmed that milk, particularly skimmed milk, was positively associated with acne. This is possibly due to higher IGF-1 levels and a lack of fat to slow hormonal absorption [270].

Dairy is one of the most common **food triggers** for eczema, particularly in children. The proteins in cow's milk can trigger an **immune response,** leading to **inflammation, itching and dry, scaly skin.** Some individuals with eczema may have **cow's milk protein intolerance (CMPI)** or **cow's milk allergy (CMA).** This is where the immune system reacts to **casein or whey proteins** in dairy.

Dairy can also aggravate **rosacea**, a chronic inflammatory skin condition that causes **facial redness, visible blood vessels and bumps. Histamines and inflammatory responses** from dairy products can **worsen skin flushing** and irritation. Research on the connection between dairy and rosacea is less extensive than for acne and eczema. But **many dermatologists** tell those with rosacea and inflammatory skin disorders to avoid dairy to see if symptoms improve [271].

Emerging research suggests that **gut inflammation** plays a key role in skin conditions. Dairy can contribute to **gut permeability ("leaky gut").** This can cause **systemic inflammation**, worsening acne, eczema and rosacea. People who cannot **digest lactose** may develop **gut inflammation** that manifests in the skin. But **casein and whey proteins** in dairy can trigger **inflammatory immune responses**, even in those who are not allergic [272].

Improving Health for Black Women

"A review of dairy food intake for improving health for Black women in the US during pregnancy, foetal development, and lactation" is a troubling paper. It examines the role of dairy consumption in supporting maternal and foetal health among Black women. The analysis was completed by looking at the results of previous studies

[273]. It highlighted that many Black women in the U.S. do not meet the recommended three daily servings of dairy. It discussed this as a possible reason for "potential" deficiencies in calcium and vitamin D. The authors of the study proposed a solution. They suggested that misconceptions about lactose intolerance need addressing. They also offered that promoting adequate dairy intake could benefit prenatal and postpartum health.

"THIS WORK WAS SUPPORTED BY AN UNRESTRICTED EDUCATIONAL GRANT PROVIDED BY THE NATIONAL DAIRY COUNCIL."

The study acknowledges that many Black women fall short of dairy intake recommendations. That this is partly due to concerns about lactose intolerance, of which three-fourths of black Americans suffer [274]. They acknowledged that black Americans are twice as likely to suffer from food insecurity, compared to the national average. The paper does not recognise the varying quality of healthcare across racial groups in the United States of America. The role that this may play in increased disease rates in black populations is not addressed. It is not noted that dairy intake, particularly in Western countries, increases with wealth and that increased wealth correlates with improved health outcomes. There is also no mention that darker skin tones synthesise vitamin D at a lower rate. Low vitamin D can lead to deficiency, impacting calcium levels. Instead, it emphasises addressing misconceptions about lactose intolerance over advising women on how to get the nutrients that they need for good health with a varied diet. The reason for this becomes clearer when looking at the funding source - the National Dairy Council. This could be considered a conflict of interest. The authors may feel the influence of the "unrestricted" funding and come to conclusions that would be positive for the National Dairy Council. The paper could appear to be a project of market research, rather than science, as the National

Dairy Council seeks new population groups to sell to. These possible conflicts of interest highlight the need for impartial funding sources for scientific projects, with an aim to come to more unbiased conclusions.

It is objectively true that there is a serious and significant disparity in the health outcomes between different racial groups in the U.S., where the paper was written. The quality of nutrition available to people across the country suffers greatly in poorer communities. It is understandable that the dairy industry would see this as an opportunity to increase sales. That black people, black women in this case and their children in the future, are an untapped market. However, the ethics of this are questionable.

DECLARATION OF COMPETING INTEREST

K.B.C. WORKS FOR THE CALIFORNIA DAIRY RESEARCH FOUNDATION. HE IS ALSO THE FOUNDER OF OMNI NUTRITION SCIENCES AND COLLECTED CONSULTING FEES FOR HIS WORK ON THIS MANUSCRIPT.

The Weinzirl Anaerobic Spore Test

By the 1910s, the public was becoming aware of the dangers of contamination of milk. This left scientists trying to find simple, cheap and quick methods for determining the safety of dairy products. The hope was that guaranteed safety would reassure the public. One hundred years ago, we did not have technology that was able to test for all possible contaminants. At the time, the concern was bacteria; today it includes medications, pathogens and faecal matter. Dipstick and rapid testing that would, within minutes, give a positive or negative result had not yet been developed. Testing at that time could be imprecise and give false results if contamination levels were low. With mass spectrometry and other technologies, it is harder to miss impurities today. But a paper, published in 1914, outlined a possible method for identifying manure-contaminated milk. The Weinzirl Anaerobic Spor Test was first outlined in the paper, A Bacteriological Method for Determining Manual Pollution of Milk, Weinzirl et al., 1914. It uses a simple method for detecting dirt contamination in milk samples [275] [276]. It starts with adding a quantity of milk to a sterile test tube and adding sterile paraffin:

> *To secure aerobosis sufficient sterile paraffin was added to the tube to make a layer one-eighth of an inch or more in thickness. The tube containing the sample of milk and some paraffin was heated to 80°C for 10 minutes, cooled, and incubated. If the sporogenes was present the lactose was digested under the anaerobic conditions and the gas formed raised the paraffin plug some distance up the tube.*

Methods such as this continued to develop, even after the milk processor introduced pasteurisation. They became a required part of the process of manufacturing milk and milk products. This testing is necessary as, at the time of milking, raw milk can contain dangerous contaminants. These include **bacteria, viruses, fungi, manure, dead skin, pus and antibiotic residues**. Without pasteurisation, raw milk **poses a serious health risk**. Many of these pathogens cause **foodborne illness, infections and long-term health issues**. But even with pasteurisation, milk is not necessarily safe, as toxins, antibiotics, pesticides and chemical residues can remain.

Veganism and the Shift Away from Dairy

"Come over to the dark chocolate side." –
Mic The Vegan

Today, milk consumption is already declining in some regions. As more people learn about lactose intolerance, animal welfare, and environmental issues, they are choosing alternatives. The dairy industry is now scrutinised for its environmental impact. This includes methane emissions from cows, high water usage, and habitat destruction. This is leading the dairy industry to go on the offensive, attacking dairy-free products. They've launched marketing campaigns highlighting the nutrient content of milk. They tout the happiness of the cows that they use.

Meanwhile, plant-based alternatives like almond, soy and oat milk are gaining popularity. They offer variety while being more sustainable and ethical. These plant milks have a much lower environmental footprint. They can provide similar or even superior nutritional benefits without the environmental costs. Oat milk, being a source of beta-glucans, is an example.

Figure 12 below shows that plant-based dairy alternatives result in lower emissions. They cause less eutrophication and use significantly less water, land and energy compared to traditional dairy products. For example, producing a litre of almond milk requires **33.6% of the land** needed to produce the same amount of cow's milk. It also generates **17.7% per kg of emissions. Eutrophication per kg of almond milk is 36.2%** of that of cow's milk. Almond milk is not a perfect solution, though. Almond milk requires less water per litre than cow's milk: **371 litres of water for almond milk, and 1,050 litres for cow's milk**. But almonds do require large water withdrawals. They are a water-intensive crop with a heavy reliance on irrigation. In cattle farming, farmers can minimise freshwater inputs using rainwater.

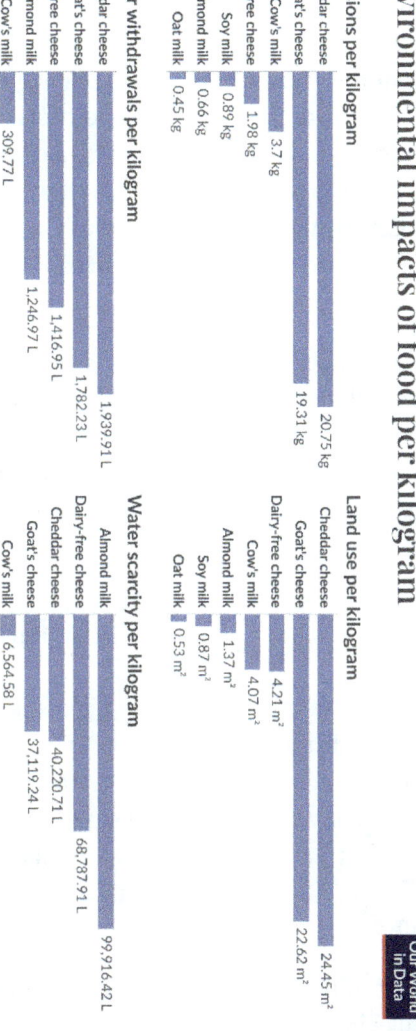

Figure 12

As a whole, almond milk is better for the planet than cow's milk, resulting in less pollution. We should note that almonds don't produce manure. Growing almond trees can also **sequester carbon**, whereas dairy farming is a net contributor to greenhouse gas (GHG) emissions. Almond trees, like all trees, absorb **carbon dioxide (CO_2) from the atmosphere** during photosynthesis. It is then stored in their biomass (trunks, branches and roots) and soil. This makes almond orchards a **carbon sink** to some extent, particularly if growers maintain the trees for long periods. A **2019 life cycle assessment (LCA)** by the **Almond Board of California** found that **almond trees store more CO_2 than is emitted during almond production**. This makes them a **net carbon sink** under sustainable farming practices.

In the context of plant milk, almonds are an outlier. Both soya and oat milk need comparably tiny inputs of water and land and result in comparably tiny carbon footprints. These alternatives also avoid the pollution and resource depletion associated with dairy farming. They are an easy choice for consumers concerned about the environment.

Nutrition

One of the main arguments against swapping dairy for plant-based dairy substitutes is the lack of nutrients in dairy-free foods and drinks. In the paper, *Plant-Based Formulas and Liquid Feedings for Infants and Toddlers,* Vandenplas et al. 2021, the nutritional value of different infant and toddler formulas was examined and compared [277]. The paper concluded that these products are all healthful and safe to use.

Plant milks also have nutritional values that cow's milk, or animal milks in general, do not have. Plants make many compounds that are of benefit to our health, such as antioxidants. Even though animals eat these compounds in their feed, they are not passed on through their milk or flesh. To get these compounds, we must eat the plant ourselves. Plant milks that keep these chemicals make consuming them easy:

Phytonutrients and Antioxidants

1. **Isoflavones (Soy Milk)** - Soy milk is rich in isoflavones. This phytoestrogen is linked to heart health and potential anti-cancer benefits. Cow's milk does not contain isoflavones but instead contains bovine oestrogens that have been connected to the promotion of cancer growth.
2. **Lignans (Flax & Oat Milk)** - Lignans are plant compounds with antioxidant and oestrogenic properties, supporting hormonal balance and heart health.
3. **Phenolic Compounds (Almond, Oat and Rice Milk)** - These have antioxidant and anti-inflammatory properties that may reduce oxidative stress.
4. **Beta-Glucans (Oat Milk)** - A type of soluble fibre that helps lower cholesterol and supports gut health.
5. **Flavonoids (Cashew & Hemp Milk)** - Natural antioxidants with anti-inflammatory and immune-supporting benefits.

Healthy Fats and Fatty Acids

1. **Omega-3 Fatty Acids (Flax, Hemp and Walnut Milk)** - These are important for brain and heart health. Cow's milk does not contain significant omega-3s.
2. **Monounsaturated and Polyunsaturated Fats (Almond, Cashew and Walnut Milk)** - These plant-based fats are beneficial for heart health and inflammation. This is the opposite of the saturated fats in cow's milk, which are proinflammatory.
3. **Medium-Chain Triglycerides (Coconut Milk)** - These fats are quickly absorbed, providing an energy boost and supporting metabolism.

Fibre and Prebiotics

1. **Dietary Fibre (Oat, Coconut and Almond Milk)** - Unlike cow's milk, many plant-based milks contain some fibre, which aids digestion and gut health.
2. **Prebiotics (Oat and Soy Milk)** - These feed beneficial gut bacteria, supporting the microbiome.

Unique Proteins and Amino Acids

1. **Edestin (Hemp Milk)** - A highly digestible protein unique to hemp, supporting immune function.
2. **Arginine (Hemp and Soy Milk)** - An amino acid that promotes circulation and heart health.

Vitamins and Minerals Not in Cow's Milk

1. **Vitamin E (Almond, Cashew and Sunflower Seed Milk)** - A potent antioxidant for skin and immune health.
2. **Magnesium (Hemp, Oat and Almond Milk)** - Essential for muscle function and bone health, but low in cow's milk.
3. **Selenium (Brazil Nut Milk)** - Supports thyroid function and immune defence.

Conclusion

Dairy farming, while often overlooked in discussions about environmental sustainability, has significant impacts on local ecosystems, global climate and natural resources. The resource-intensive nature of dairy production combines with the pollution and habitat destruction it causes. This makes it a major contributor to environmental degradation.

The production of dairy involves the exploitation of cows. It turns their bodies into baby-making, milk-producing machines. Cows bred to produce huge quantities of milk suffer from pain as their udders are stretched beyond what is natural. They will wait, in severe discomfort, for their turn to be milked. If these cows kept their calves, the milk would continually feed the calves, preventing this overload on their bodies. The lives of these cows are stressful. They live by the will of the farmer, by the will of the industry. They are forced to breed only to have their calf taken away. This is not natural, it is not necessary, and it is not healthy.

In assessing the environmental impact of dairy farming, the evidence is clear. It is a massive contributor to greenhouse gas emissions, deforestation, water consumption and pollution. It is a massive driver of climate change. While policy changes and technological improvements may reduce some of these impacts, they do not eliminate them. Supplementing cow feed to reduce methane in burps does little. Making a cow where a mask is not the answer. Shifting away from dairy presents a direct and effective way to lower emissions, conserve resources and reduce environmental degradation. With so many plant-based alternatives for sale today, the shift is easy. The connection between dairy and climate change is well-documented. The choice to act on this knowledge is available. To actually make the change is both a practical and necessary step toward a more sustainable future.

Chapter 9: Wool, Feathers, Skins and Fur

Introduction to Animal-Derived Fabrics

Clothing and fabrics made from animal skins, fur and hair have been a staple of human civilisation for centuries. These materials are often associated with luxury, warmth and durability. But they come with significant costs, particularly to the environment. The production of these materials involves direct harm to animals. But it also causes substantial environmental impacts. Here again, we find pollution, resource depletion and habitat destruction.

Fashion is one of the most polluting industries in the world. It uses many non-renewable materials and huge amounts of water. Dangerous chemicals are a part of many manufacturing processes. The fashion industry also results in vast amounts of waste going to landfills. Landfills contribute to air, land and groundwater pollution. There are many figures given for the damage that the fashion industry causes. These include carbon emissions and water use. People measure how much clothing becomes waste, with different studies producing different results. Moorhouse conducted a study on this in 2020. The paper, Making Fashion Sustainable: Waste and Collective Responsibility investigated processes of fashion waste disposal. It found that 73% of clothing, both new and used, ends up in landfills, and less than 1% is recycled into new clothing [278]. Fashion, including clothing, shoes and apparel, requires large inputs of fossil fuels. The manufacturing processes are long, using chemicals, huge machinery and transportation of products. 1.2 billion metric tons of CO_2 were reportedly emitted by the fashion sector alone in 2015. Textile dyeing and finishing contribute to roughly 20% of the world's water pollution [279]. Manufacturing, dyeing and processing consume vast volumes of water.

When using animals and other living creatures, such as insects, to make fabrics and clothing, we see familiar patterns. Environmental damage, pollution and exploitation. The same negative impacts that we see in the farming of animals for meat, fish, eggs and dairy. People take huge areas of land to farm animals and grow feed. Growing feed, watering animals and processing the products that come from them consume millions of litres of water. Animal waste and chemical outputs pollute the air and water. Soil becomes degraded, made hard, infertile and barren. Soil ends up being impenetrable to water and the roots of plants.

Silk: The Environmental and Ethical Costs

The History of Silk

Valued for thousands of years, the origins of silk trace back to ancient China. There is no documented history of the origin of silk fabric. But according to legend, Empress Leizu discovered silk around 2700 BCE, when a silkworm cocoon fell into her tea. The cocoon unravelled into a fine, strong thread that was used to make fabric. Regardless of the origin, the Chinese created **sericulture**, the practice of raising silkworms and harvesting their silk. For centuries, they maintained a monopoly on its production. This established the **Silk Road** during the Han Dynasty (206-220 CE). This was a trade route between China, the Middle East and Europe. The buying and selling of silk made it a prized commodity.

By the Middle Ages, silk production had spread beyond China. Countries like Japan, India and parts of the Middle East started using sericulture. European countries, particularly Italy and France, became major producers by the Renaissance. They refined weaving techniques and developed luxury silk goods. However, the industry remained reliant on labour-intensive processes. These included cultivating silkworms and extracting silk by boiling or steaming cocoons, killing the larvae inside.

Technological advancements in the 20th century brought changes to the silk trade. Synthetic fibres like nylon and polyester. They provided cheaper alternatives, making silk less attractive to consumers. Despite this, demand for natural silk persisted, particularly for high-end fashion and luxury textiles. In response, silk production expanded into new regions. In 2023, countries like China, India, and Brazil produced almost 94,000 metric tonnes of silk [280].

The Process

Most commercial silk production still follows traditional sericulture methods. Modern technology has only improved efficiency. The process begins with raising silkworms, primarily the domesticated species *Bombyx mori*. The *Bombyx mori* is specifically bred for silk production. Over thousands of years of selective breeding, this domestic breed has lost many of the traits needed for survival in the wild. It is larger than the natural silk moth, and it is white, having lost its natural brown camouflage. The Bombyx *mori* produces more silk than the wild breed, and it cannot fly. They have been so overbred that they can no longer find food or even reproduce without human help. Their entire lifecycle depends on human intervention. These weaknesses make them one of the most domesticated insect species in the world.

The caterpillars of the silk moth only feed on mulberry leaves. After 25-30 days, they spend 3-4 days spinning cocoons made of a single continuous silk thread. This thread can be up to 900 meters long.

Once the silkworms have completed spinning, workers harvest the cocoons. Most commercial producers soften cocoons by steaming or boiling. This process kills the pupae inside, but allows them to extract long, unbroken silk fibres. Killing the pupae prevents moths from emerging and breaking the silk thread. The softened cocoons are then unwound, most often by hand. This part of the process takes hours; one person can pull around 70g of silk per day. The thread is then spun with several strands twisted together to create a silk thread. This thread is then dyed, woven into fabric and finished for use in textiles. It takes roughly 3000 cocoons of thread to make a silk dress.

Silk also has uses outside of fashion, such as in medicine and the beauty industry. In healthcare, silk protein can make dressings. For medication delivery products, pharmaceutical companies use silk in slow-release capsules. Silk threads are still used in some procedures for sutures and as a form of scaffolding for tissue engineering. The properties of silk make it ideal for regenerating skin, bones, nerves and cartilage. Cosmetic and beauty products use silk as an ingredient for hydration and skin repair.

The Environmental Impact

People value silk for its softness and lustrous appearance. But its production has both environmental and ethical implications. It is a resource-dependent fabric, needing large amounts of land and water for production. It also relies on the planting of mulberry trees to produce food for the silkworms.

Resource Use in Silk Production

Land: Silk production needs a lot of land. This land is used to grow mulberry trees, which feed the silkworms. The expansion of mulberry plantations can contribute to deforestation and habitat destruction. Silk production typically occurs in poorer regions of the world. In these regions, people take advantage of natural resources, clearing land to make way for silk production. Mulberry trees, planted for feed, grow in monocrops. This causes the same ecological damage as growing corn and soya for other animals. Silk farmers clear wildland, replacing native forests with monocrop mulberry crops. The process reduces biodiversity and disrupts local ecosystems. The loss of natural habitats affects wildlife. It damages soil health, leading to erosion and reduced carbon sequestration.

The mass breeding of domesticated silkworms has led to fewer wild silkworm species. This decline reduces genetic diversity in silk-producing moth populations.

Water: Silk production is also highly water-intensive, impacting both water availability and quality. Mulberry trees need a lot of water. This means farming them requires significant irrigation. This is particularly true in regions with dry climates, causing strain on local water resources. Producing one kilogram of silk fabric can use up to 5,000 litres of water, making it one of the most water-intensive textiles.

If we go back to the example of the silk dress, we can see how resource-intensive silk is:

A Silk Dress: A typical silk dress requires about **450-600 grams (1-1.3 lbs)** of raw silk, depending on the style and fabric thickness.

Mulberry Leaves Required: Silkworms feed exclusively on mulberry leaves. To produce **600 grams of silk**, approximately **30-40 kg (66-88 lbs)** of mulberry leaves are needed.

Number of Cocoons: Each silkworm cocoon produces up to **900 meters** of silk filament, but only **about 450-600 meters** is usable. It takes around **2,500-3,000 cocoons** to produce enough silk for one dress.

Land Requirement: Mulberry trees need approximately 1 acre (0.4 hectares) of land to produce 500-700 kg (1,100-1,540 lbs) of leaves per year. This can support several silk harvests. A single silk dress requires about 40-50 square meters (430-540 square feet) of land for the mulberry trees.

Water Consumption: The production of silk is water-intensive. This is because water is used for both mulberry cultivation and silk processing. Estimates suggest that a silk dress would need **3,000-5,000 litres (790-1,320 gallons) of water** to produce. This includes water for irrigation, silkworm rearing and processing (such as degumming and dyeing).

Resource Usage for One Silk Dress
- **Silk required:** 450-600 grams (1-1.3 lbs)
- **Mulberry leaves:** 30-40 kg (66-88 lbs)
- **Cocoons used:** 2,500-3,000
- **Land needed:** 40-50 m² (430-540 ft²)
- **Water required:** 3,000-5,000 l

Sources:
CFDA.com [281]
Fritz et al. 1986 [282]
fabricmaterialguide.com [283]

The Environmental Pollution of Silk

Greenhouse Gases: Many believe that silk is a natural and biodegradable material. However, it has a surprisingly high environmental footprint. The process of making silk produces a lot of pollutants, impacting air and freshwater sources. One of the primary sources of greenhouse gas emissions is growing mulberry trees for silkworms. The process adds to **carbon dioxide (CO_2)** emissions through land clearing, soil degradation and agricultural machinery use. Additionally, nitrogen-based fertilisers release **nitrous oxide (N_2O).** This greenhouse gas has nearly **300 times** the warming potential of CO_2.

The energy-intensive process of boiling or steaming cocoons to extract silk threads also generates emissions. Growers need temperature-controlled environments to breed silkworms, with 22°C being optimal. In temperatures greater than this, air conditioning regulates the atmosphere for the insects, as higher temperatures will lead to death. This is common in regions of China, India and Brazil where most silk production takes place. Silk processing facilities rely on fossil fuels such as coal and natural gas to power their operations, adding to CO_2 emissions. Furthermore, traditional silk production is less efficient compared to other textiles. Studies have found that silk has one of the highest carbon footprints per kilogram of fabric. The emissions surpass those of even cotton and synthetic materials.

Air Pollution and Toxic Chemical Use

Dyeing and finishing processes of silk release **volatile organic compounds (VOCs)** and hazardous chemicals into the environment. Bleach, synthetic dyes and finishing agents, used to treat silk fabrics, release pollutants into the air and water. In countries with lax environmental regulations, untreated industrial waste from silk factories causes air pollution. Workers and nearby residents in these countries suffer from exposure to toxic substances.

Water: Silk production has a large water footprint. But the required processing also generates chemically polluted wastewater. Raw silk thread, from silkworm cocoons, has a protein coating, sericin. Silkworms secrete this protein to keep the cocoon intact and protect the fibres. Degumming - the removal of the sericin - uses chemical treatments that produce wastewater containing detergents and heavy metals.

Silkworms are also prone to diseases and parasites when farmed. Keeping the larvae in containers allows diseases and infections to spread rapidly. Treatment for these illnesses involves the use of disinfectants, including bleach and lime. Sometimes the only treatment is to burn the bodies of infected larvae. During treatments, workers wash disinfectants, normally untreated, into the environment. These chemicals are toxic to waterways and the surrounding area. The releases can contaminate rivers and groundwater. They harm both aquatic ecosystems and communities that rely on these water sources.

Overuse of insecticides and pesticides in mulberry farming adds to water, air and soil pollution. Many farmers rely on chemical treatments to protect mulberry crops from pests. This leads to long-term environmental destruction and biodiversity loss. Excessive use of pesticides also creates health risks for workers while contaminating local ecosystems.

The Need for Sustainable Alternatives

The environmental effects of silk have made people more interested in **sustainable and ethical** options. Lab-grown silk, plant-based fibres and recycled silk fabrics are all good alternatives. They reduce pollution while maintaining the qualities that make silk desirable. Innovations in **closed-loop water systems** and **eco-friendly dyeing processes** are also being developed. These processes mitigate the environmental damage caused by traditional silk production. Public awareness of the damage that silk production causes is pushing the industry to change. People want cleaner, less resource-intensive fabrics with a reduced impact on the planet.

Some producers use alternative methods, such as **Ahimsa silk** (peace silk). This method allows moths to emerge before harvesting the silk. However, this results in shorter, broken fibres that need spinning rather than reeling. It also creates a fabric with a rougher texture. This method still needs land for the monocrop growing of mulberry trees. It still uses excessive amounts of water, and the method does not prevent any pollution or the release of greenhouse gases. Other innovations use lab-grown and plant-based silk. These aim to replicate silk's properties without depending on animal farming. These alternatives offer a more sustainable and cruelty-free option for concerned consumers.

Wool: The Environmental Impact of Sheep Farming

Wool, prized for its warmth, softness and its moisture-wicking properties, is popular for use in fashion. But is also used by those practising extreme sports in cold environments. Many consider wool to be a renewable material. But, its production, including scouring and dyeing, has significant environmental implications, as do the impacts of sheep farming before wool is harvested.

The History of Wool

The history of wool dates back thousands of years. Humans began domesticating sheep for their wool as early as 6000 BCE in Mesopotamia after they first raised them for meat. Once people realised how useful wool could be, they began selective breeding. They developed sheep with thicker, more productive fleeces. Early civilisations, including the Sumerians and Egyptians, also used wool for clothing. By 3000 BCE, wool production was well-established in regions such as Persia, Greece and Rome. It was these societies that developed methods for shearing, spinning and weaving wool, making it one of the most valuable textiles in ancient times.

During the Middle Ages, wool production flourished, particularly in Europe. Countries like England, Spain and Italy became major centres of the wool trade. England's economy was especially reliant on wool exports. The English monarchy even imposed strict laws to protect the wool industry. Meanwhile, Spain developed the Merino sheep breed. This breed produces a soft, fine wool that was initially only used by Spanish royalty. Later, as supply increased, worldwide trading began. Merino sheep were also valuable as they did not shed their fleeces. Farmers would leave their fleeces to grow longer before shearing. This is different to wild breeds that moult seasonally.

With the Industrial Revolution in the 18th and 19th centuries, wool production became more efficient due to mechanised spinning and weaving. This led to a massive expansion of the wool industry, particularly in Britain and its colonies. This included Australia and New Zealand, where sheep farming began with 13 animals in 1797. Today in Australia, sheep far outnumber humans. In 2022, 70.2 million sheep produced 350 million kilograms of wool in this country (284).

Sheep Farming

Farming sheep for wool, as with the farming of other animals, begins with breeding. Farmers manage this with artificial insemination or controlled natural mating. Artificial insemination in large-scale operations maintains wool consistency. As in beef and dairy farming, it allows farmers to select specific traits from rams. Farmers choose qualities like high wool yield or finer fibre to increase their profits. Ewes, female sheep, will have their first pregnancy at around **8 months old** with a gestation period of about five months.

Lambs will often have their tail docked. It is a practice claimed to reduce the risk of flystrike. Male lambs may also be castrated. Farmers do this using either a hot knife or a rubber band. The tight band restricts blood flow to the testicles until they die and fall off. These procedures are carried out without pain relief. In some wool-producing regions, particularly in Australia, mulesing prevents flystrike on Merino sheep. Flystrike is an infestation where flies lay eggs in the flesh of live sheep. The maggots then hatch on the skin of the living animals. The museling procedure involves cutting away skin around the sheep's rear. This is another procedure carried out without anaesthesia, leaving sheep with an open wound, which causes significant pain.

Farmers raise lambs in pastures or confined feedlots, depending on the farming system. Sheep bred for wool are shorn once or twice a year. Merino sheep need shearing more often because their fleece grows continuously. Shearing is a stressful process for the animals. It is often done as quickly as possible to maximise efficiency, leading to cuts and injuries.

Throughout their lives, wool sheep are repeatedly shorn. But their usefulness in the industry declines as they age and wool quality reduces. Once farmers no longer find them profitable for wool production, they send them to slaughter. This will happen when sheep are 5-7 years of age. By this point, their bodies start to suffer from the stress caused by farming, whereas in the wild, sheep can live to 20 years of age or more.

In many cases, sheep travel long distances to abattoirs, an exhausting and stressful journey. In some countries, sheep travel by live export. In this system, sheep travel for weeks in crowded, unsanitary conditions before reaching slaughterhouses overseas. At the slaughterhouse, sheep are stunned - either electrically or using a captive bolt - before the killing. When stunning methods are not effective, animals remain conscious during slaughter. This happens far too often, although it should never happen at all.

The Process

Rearing

The production of wool begins with sheep rearing. Farmers raise sheep specifically for their fleece. Most wool-producing sheep breeds, such as Merino, have thick coats that grow continuously, allowing regular shearing. These sheep are often kept in large flocks on pastures or in intensive farming systems. Farmers manage the sheep to maximise wool growth, feeding them a specialised diet. Intensive sheep farming, though, can lead to environmental issues such as overgrazing, soil erosion and water pollution.

Shearing

Shearing is the next step in wool production. This involves using electric clippers or manual shears to remove the fleece from the sheep. This is usually done once a year in spring or summer. The shearing process is stressful for the animals, and in large-scale operations, it is done quickly, sometimes resulting in injuries. Shearers receive payment based on the number of fleeces they collect, rather than the time they spend. For this reason, shearing is often rushed, with workers wanting to collect at least 140-160 fleeces per day, which is the average for an experienced shearer.

Shearing is especially stressful for sheep. It happens only once or twice a year, meaning that sheep don't get used to the procedure. But also, sheep are prey animals. Being held down, put on their backs and manipulated while someone cuts their fleece away creates a prey response in the animals. A fear response. They will attempt to herd, hide in the group, or run away. They will eventually freeze as they realise there is no escape.

Production

After collection, the fleece goes through cleaning and processing. Raw wool, also known as grease wool, contains dirt, sweat, plant debris and lanolin, a natural oil. To remove these impurities, the wool first has the most flawed areas of fleece removed by hand. This includes pulling out clumps with a lot of plant matter or manure contamination. The fleece is then washed in a process called scouring. The process involves soaking the fleece in hot water with detergents. Chemicals used can include surfactants or even sulphuric acid. Scouring washes and rinses the fleece multiple times. The fleeces go into huge, industrial-sized machines that carry the fleeces from one bath to the next. After cleaning, the wool is dried in industrial dryers before scattering. This is a process of spinning the wool into the air, separating the fibres. The wool is then collected and pressed into bales. Brushing aligns the fibres, preparing them for dying and spinning, which is called carding.

Wool is then made into yarn and dyed or dyed first. Dying can use natural dyes, but more often uses synthetic dyes, for reasons of cost. Spinning twists the wool fibres together to create yarn. Once dyers and weavers process the wool, they use it to make garments, carpets, and other woollen products. The final stage involves finishing processes such as shrinking, softening and anti-pilling treatments. These enhance the quality of the wool fabric, often using chemical processing.

The Environmental Impact

Resource Use

Land Use and Deforestation: **Land Use and Deforestation**: The farming of sheep is very land-intensive. It uses the most land per 100g of protein of any food source at 185 m². In the production of wool, this use of land is inefficient in the extreme. Researchers estimate that approximately 2% of the energy in the feed ends up in the wool. This means that for every 100 units of energy a sheep consumes, roughly 2 units become wool. The remaining 98% keeps the animal alive, while it loses a lot as heat or other by-products (such as manure and methane) (285).

Unlike other forms of animal agriculture, sheep graze and forage for their food. Farmers use less than 10% of the land used in sheep farming for growing crops. The method of grazing means that sheep have access to many acres of open land, land that could be used more efficiently to grow crops for humans. Overgrazing by sheep on these vast areas of terrain can cause **soil degradation, desertification and reduced land productivity**. Intensive sheep farming also leads to vegetation loss. This causes more erosion, less water retention, and harms ecosystems. The loss of soil, soil that can support life, including plants, insects and fungi, is catastrophic for life on Earth.

Soil is a complex mixture of minerals, organic matter, water, air and living organisms. They all work together to create a thriving ecosystem. Soil has three main mineral components: sand, silt, and clay. It also contains organic matter from decayed plants and animals. Together, these elements give plants the nutrients and stability they need to grow. This, in turn, supports entire food chains.

Healthy soil is the foundation of life on land. It plays a crucial role in water filtration, helping to purify rainwater as it percolates through the ground. It acts as a carbon sink, storing vast amounts of carbon and helps to regulate the Earth's climate. Without life-sustaining soil, ecosystems would collapse, and agriculture would become impossible. Biodiversity would be harmed to the point of no recovery.

Sheep farming destroys soil health, with overgrazing being a major issue. Sheep can strip the land of protective plant cover through continuous feeding. This kills the roots of plants that bind the soil, allowing wind and rain to wash or blow away the nutrient-rich topsoil. This can lead to desertification, making the land less productive over time. Overgrazed pastures also become compacted, reducing the soil's ability to absorb water. This increases runoff as manure sits on top of the soil and contributes to drought conditions. In this state, plants cannot penetrate the soil with their roots. Air pockets are squashed out. The delicate balance of soil microorganisms becomes disrupted, further reducing soil fertility. In wetter regions, sheep can exacerbate problems like waterlogging and increased methane emissions from disturbed soil.

The most effective way to protect soil from the damage caused by sheep farming is not to farm them. After that, it is to transition away from land-intensive livestock farming and move toward plant-based agriculture. This type of farming uses less land and preserves soil health.

Water Use: The wool industry requires large quantities of water. Substantial water consumption occurs at several stages of production, from sheep farming to processing the final fabric. One of the largest water demands comes from the raising of sheep. They need drinking water and irrigation to maintain pastureland in drier areas. Soil damage also encourages drought conditions in water-stressed regions when the ground loses the ability to absorb water.

As well as wool, people farm sheep for milk and meat. Farmers slaughter sheep when they are no longer profitable for milk or wool, or when they have reached slaughter weight if raised for meat. In Australia, sheep meat (mutton) is popular, with people eating on average 6 kg of mutton each year. Australia is also the world's top exporter of mutton, lamb and live sheep. Using a lot of the country's freshwater resources [286].

Sheep slaughter follows a standardised process similar to that of cows. Animals are first **stunned** to render them unconscious. Slaughterhouse workers use either **electrical stunning** or **captive bolt stunning**. Electrical stunning uses electrodes applied to the head to induce a seizure. Captive bolt stunning shoots a metal bolt into the skull of sheep. Religious slaughter (such as halal or kosher methods) can skip stunning. Once stunned (or restrained in religious slaughter), workers cut the sheep's throat (**exsanguination**). They sever the carotid arteries and jugular veins, causing rapid blood loss and death. The body is then processed by removing the fleece, internal organs and head before preparing the meat for distribution.

The total water footprint for the process of raising, slaughtering and processing sheep is high. It includes **water for drinking, feed production, cleaning and slaughterhouse operations.**

1. Water for Raising Sheep

- **Drinking Water:** A single sheep typically drinks **3-12 litres (0.8-3.2 gallons) per day**, depending on age, size, climate and diet.
- **Water for Feed Production:** Most water use in sheep farming comes from growing feed (grass, hay and grains), ranging from **several hundred to thousands of litres per kilogram of feed**.

2. Water in Slaughter and Processing

- Slaughterhouses use water for stunning, bleeding, carcass washing and equipment cleaning. On average, **250-550 litres (66-145 gallons) of water** are used per sheep in slaughterhouse operations.

Total Water Footprint per Sheep

The total **lifetime water footprint** of a sheep is estimated to be around **5,500-10,000 litres (1,450-2,640 gallons) per kilogram of meat**. Since a sheep typically yields about **20-30 kg (44-66 lbs) of**

meat, the total water use per animal can exceed **100,000-300,000 litres (26,000-79,000 gallons).**

Sources:

Mekonnen, M. M., & Hoekstra, A. Y. (2010) [60]

Mekonnen, M. M., & Hoekstra, A. Y. (2012) [61]

FAO (2006) [287]

Willett, W. et al. (2019) [288]

The processing of wool after shearing is a major source of water consumption. Scouring, the washing process used to remove dirt, grease, dead skin and sweat from raw wool, requires large amounts of water and detergents. The baths used, as fleeces pass from one vat of hot water to the next, can contain as much as 2000 litres of water. To heat each bath from cold each day would be time and energy-consuming as well as extremely expensive. Dyeing wool also takes place in large volume baths, requiring many thousands of litres of water again, to colour, wash and rinse the fibres.

The total water footprint of producing a woollen jumper, from beginning to end, is massive. This is starting with sheep rearing to processing the final fabric. Estimates say that producing one kilogram of clean wool requires **tens of thousands of litres of water**. Let's break this down step by step:

1. Sheep Rearing (Drinking Water + Feed Production)

- A single sheep drinks **6-12 litres of water per day** (depending on climate and size).
- Sheep raised for wool live around **5-7 years** and are sheared annually.
- Wool yield per sheep varies by breed, but on average, a Merino sheep produces **4-5 kg of raw wool per year**, which results in **about 2 kg of clean wool** after scouring.

- Growing feed and maintaining pastureland for one sheep requires around **50,000 litres of water per year** in some cases.
- Based on these numbers, the water footprint per kilogram of clean wool from sheep rearing alone is estimated at **50,000-100,000 litres**.

2. Wool Processing (Scouring, Carding, Spinning, Dyeing, Finishing)

- Scouring (washing raw wool): Requires 100 - 400 litres of water per kilogram of wool, depending on the method and efficiency of the facility.
- **Dyeing**: Can use up to **600 litres per kilogram** of wool fabric, especially with synthetic dyes.
- **Other processing stages (carding, spinning, finishing)**: Add several hundred litres more, bringing total wool processing water use to around **1,000-2,000 litres per kilogram**.

3. Wool Fabric to Jumper Conversion

- A standard woollen jumper weighs **about 600 grams to 1 kilogram** (depending on thickness and size).
- If we assume **1 kg of wool fabric per jumper**, then the water footprint for **one woollen jumper is roughly 150,000-200,000 litres.** This assumes maximum water inputs for all stages (sheep rearing, feed, scouring, dyeing and finishing). More conservative estimates calculate the water footprint to be around **77,000 litres**. The figures vary by sheep breed, climate and farming methods. Regardless, 77,000 litres of water for a jumper is not only extreme but unnecessary.

Fossil Fuels: Apart from niche and artisan markets, every part of wool production happens on an industrial scale. In European countries, the number of sheep on a farm can number a few hundred. In New Zealand, sheep flocks can range from 1000-5000 animals. In Australia, flocks can number up to 20,000. When farming sheep for wool, fossil fuels are used in every stage of the process. Growing crops for feed requires fossil fuels to produce synthetic fertilisers. Farm machinery is powered by fossil fuels, as are vehicles used to transport animals, feed and equipment. Processing of wool has the greatest fossil fuel inputs to power the machines that wash, dry, brush and spin the wool. Wool processors must maintain water at temperatures greater than 80°C. Water must be this hot for wool washing processes to be effective. These wool processing factories operate 24 hours a day, with staff working a shift pattern. Powering these large factories uses a lot of fuel. So, it's cheaper to keep the machines running. This means heating hundreds of thousands of litres of water all the time. If they heated water from cold each day, the costs would leave little to no profit. Also, heating thousands of litres of water to 80°C would take hours.

Environmental Pollution

Air: In animal agriculture, sheep are the second greatest emitters of methane after cattle. They produce methane through their digestive processes as they are ruminant animals. As with cattle, this is a recognised problem. Mitigating strategies farmers use include breeding sheep that have altered digestive systems. They add additives, such as olive or seed oil supplements, to feed, as well as using feed that is more processed. Sheep, classed as small ruminants along with goats, contribute approximately **475 million tons of CO_2eq to greenhouse gas emissions**. This makes up around **6.5% of world agricultural sector emissions** [289]. The methane emissions from sheep contribute to global warming. These emissions make wool production a significant source of greenhouse gases.

The production of methane in the digestive systems of animals also highlights the inefficiency of using animals to produce products, meat, milk, wool or skins. This is because methane production results in energy loss. The methane produced is energy that never makes it into the animal's body or the products that people farm them for. Studies have estimated that **2% to 15%** of the energy from feed can be lost in the form of methane. This means that a portion of the calories contained in the feed is wasted, rather than converted into biomass.

Nitrous oxide (N_2O) emissions in the wool industry mostly come from feed production at $\geq$ 35%. N_2O emissions from manure account for ≈25%, depending on the environment of the region and farming methods used. One approach for reducing methane and nitrous oxide emissions from manure includes cooling manure to below 15°C. This cooling would be another process requiring the use of fossil fuels. In mitigating one problem, you create another.

Land: In confined farming systems, such as those used for chickens or pigs, infection is a primary concern, requiring the mass use of antibiotics. In farming sheep that spend large periods grazing on open land, **infestation** is a problem.

Parasites that infect sheep can be internal, such as worms, or external, like ticks or mites. Internal parasite treatment uses oral **anthelmintics** and anti-internal parasitic medications that can expel the parasite from the body. External parasites need to be treated with **organophosphate (OP)** or **synthetic pyrethroids (SP)**. These chemicals are applied by plunge-dipping sheep in liquid insecticides and fungicides.

To treat sheep with organophosphates, they are dipped in a tank of the chemical. They must be fully submerged for at least a minute and dipped at least twice. The tanks are large, containing thousands of litres of the highly toxic liquid. The toxicity of the solutions means that people carrying out the procedure must be fully protected. These chemicals are easily absorbed through the skin; even a splash to the arm or leg is dangerous.

If anybody gets heavily contaminated, e.g. by falling in the dip, it should be treated as an emergency. Remove any contaminated clothing, wash any contamination from the skin and take the person straight to hospital.
– Health and Safety Executive [290]

Organophosphate exposure can cause coma, vomiting and diarrhoea. In the long term, it can cause cancer, issues with fertility, and brain or nerve problems [291]. Pyrethroids are safer for farmers, but they can also interfere with nerve and brain function for those exposed to the chemicals [290].

Personal Protective Equipment (PPE) required for sheep dipping

■ non-lined synthetic rubber gloves (heavy duty gauntlet-style PVC or nitrile at least 0.5 mm thick and at least 300 mm long)
■ wellington boots
■ waterproof leggings or trousers made of nitrile or PVC;
■ a waterproof coat or a bib apron made of nitrile or PVC over a boilersuit or similar;
■ a face shield.
The Health and Safety Executive [290]

Once dipped, sheep are released into an open pen. They will attempt to shake themselves dry, spraying the toxic chemical mixture onto the ground. Farmers try to keep this shaking to a minimum to prevent contamination of soil and groundwater, depending on the soil type. Sheep must also have access to drinking water for two weeks following dipping. This is to stop them from entering natural water sources, such as a stream, leading to water contamination [292].

Water: Sheep farming has the same problems of polluting waterways and groundwater with manure as pig and cattle farming. In sheep farming, the manure, containing phosphates, nitrates, E. Coli and other pathogens, is deposited onto the ground as animals graze. Runoff is a common problem, as rainwater carries the manure straight from the soil into freshwater sources.

Processing wool uses chemicals, such as detergents and acids, to clean fleeces and break down organic matter tangled in them. Industrial-scale scouring facilities use chemical-intensive methods. The resulting wastewater often contains lanolin, pesticides (from sheep dipping) and synthetic detergents. These will pollute water systems if not properly treated. In many cases, scouring wastewater is difficult to clean due to the mix of organic and chemical contaminants. The remaining chemical mixture makes it a significant environmental concern.

Wool is also rarely sold in its natural colour; instead, dyes create the yarn colours that people want to buy. These chemical dyes, normally synthetic, do not break down without the use of neutralising chemicals. The dyeing process also uses more dye than is consumed in the process. Most of the dye ends up being washed away, resulting in harmful environmental effects. Water becomes polluted with dye chemicals. This is another waste stream that requires treatment before release to prevent harm to wildlife. Natural dyes can be a better alternative, but they still need large volumes of water for extraction and application.

Sustainable Alternatives to Wool

Wool is often marketed as a sustainable and natural fibre. But its production is deeply tied to the broader sheep farming industry. This industry treats sheep as commodities rather than sentient beings. This has led to a growing interest in ethical alternatives, including plant-based and synthetic fibres, as well as calls for reform in the treatment of wool-producing sheep.

Another problem with wool that is rarely talked about is the discomfort and irritation that some people suffer when wearing the fabric. A minority of people cannot wear wool as it inflames their skin, causing itchiness and rashes. Of this group, some are allergic to wool. For many, the reaction happens when their skin comes into contact with chemical residues. These include acids used in processing fleeces and wool fibres.

Sustainable alternatives to wool include plant-based fibres such as organic cotton, hemp and bamboo. People can also choose synthetic fibres made from recycled materials. These alternatives have the benefits of wool, like warmth and moisture-wicking, without the environmental and ethical concerns.

Comparison to Other Fibres

Wool, as a fibre used to make clothing, is optional; it is something that we don't have to use. As with silk, there has already been a shift in the wool market. People are choosing the alternatives, and the demand for wool is decreasing. This change has several reasons. People are considering the cost of items made from wool. Wool is more difficult to maintain, requiring special detergents. Also, the availability of non-wool items that are cheaper and easier to care for makes them an easy choice. Some of these alternative fabrics are synthetic, such as polyester. But there are also natural fibres that have much smaller environmental impacts, like cotton and hemp.

CO₂eq: The carbon footprint of wool is higher than the alternatives that people are turning to. According to data collated by CO2 Everything, the carbon footprint of 2 square metres of wool is **13.89 kg CO₂eq**. For comparison, **cotton** has a carbon footprint of **8.3 kg CO₂eq, nylon** has, **7.31 kg CO₂eq, polyester has 6.4 kg CO₂eq,** and **linen has 4.5 kg CO₂eq**. For wool, a carbon equivalent of 13.89 per 2 square metres equates to driving for 43.8 miles, or 70.5 km. Polyester, with a carbon equivalent of 6.4 per m², would equate to driving 20.2 miles or 32.5 km. **Hemp** is an outlier in that it sequesters carbon. It removes it from the atmosphere and has a carbon footprint of **-2.48 CO₂eq per 2 m²** [293].

Water Footprints: Wool has a greater water footprint than many other fabrics. This is because you have to grow a whole animal before you can harvest the fibres that they grow. These fibres then go through processing to produce fabrics. This means that embedded in the water footprint of wool is water for drinking, growing food and general farm use, before any fibre is produced. A sheep is sheared for the first time at 6-12 months of age, depending on whether it is warm enough for the lamb to be without a fleece. During these 6-12 months, the animal is using resources that are part of the cost of the final product, both financial and environmental.

According to the analysis paper, *The Water Footprint of Cotton Consumption* (Chapagain et al, 2005), it takes 2720 litres of water to produce a cotton t-shirt [294].

If we take this number, we can calculate the water footprint of 2 m² of cotton

- A standard cotton **T-shirt** uses about **0.5 m² to 0.6 m² of fabric,** with an average of **0.55 m²** (varies by size and style). So, **2,720 litres** of water produce **0.55 m² of fabric**.

$$\frac{2720\ litres}{0.55\ m2} = 4945\ \text{litres per m2}$$

We multiply this answer by 2 to get the water footprint for 2 m² of cotton fabric: It takes **9,890 litres of water to produce 2 m² of cotton fabric**.

If we apply the same calculation to other fabrics, we get:

Water Footprint for 2 m² of Fabric

Fabric Type	Water Footprint (litres)
Wool	105,000 L
Cotton	9,890 L
Linen	807.5 L
Hemp	23.3 L
Nylon	64 L
Polyester	24 L

Table 3

Sources:

Water Footprint Network [60] [294] [295]

Alliance for European Flax-Linen and Hemp [296]

Textile Exchange [297]

Life Cycle Assessment of Linen [298]

Leather: The Hidden Costs of Animal Skins

In chapter 6, Burning Forests for Beef and chapter 8, Dairy Farming's Dirty Footprint, we explored the impacts of farming cattle. For the most part, leather is a by-product of these practices. There are situations where cows are being farmed only for leather production.

The hides of these cows produce high-value products. Regardless of the reason, the damage is the same. Air pollution, greenhouse gas emissions, high water requirements and water pollution result. There is extreme land use, deforestation and soil degradation. Still, there appears to be no end in sight to the use of these animals.

The global leather market is a lucrative one. It is worth more than $270 billion per year and is especially valuable to low-income countries as a manufacturing export. Leather is one of the most widely used animal-derived materials. It is found in everything from clothing and footwear to furniture and accessories. But even though it's considered a natural product, it is a highly processed good. Leather is the fourth most dangerous industry for human health due to its processing methods. Countries in the Global South contribute to this disproportionately, as many tanneries lack basic protection for workers or the environment, leaching toxic chromium into rivers [299].

History

The history of leather goes back thousands of years. Humans first started using animal skins for clothing, shelter and tools as far back as 7000 BCE. Early civilisations developed methods to prevent raw hides from decomposing. They would use natural processes like drying in the sun, soaking them in water and rubbing them with salt, oils, or animal fats. These early tanning techniques helped preserve the hides and make them more durable.

Tanning is the process of turning animal skins into leather, and over time, humans developed more advanced tanning techniques. These included using vegetable matter, alum salts and brain matter to make leather soft, flexible and long-lasting. Vegetable tanning (Ancient Egypt & Mesopotamia, ≈3000 BCE) used tannins from tree bark and plant matter. Ancient Egyptians used this technique to produce leather for sandals, shields and furniture. It was between 200 BCE and 400 BCE that the Romans applied aluminium-based compounds for alum tanning. The process produced a white-coloured leather used to make clothing and military gear.

Indigenous peoples across the world, including Native American and Arctic cultures, used brain tanning. They used the brains of the animals the skin came from, with water, to soften and preserve hides.

The introduction of chromium salts in the 1850s revolutionised leather production. It made the process faster and produced more uniform leather. But it came with significant environmental costs.

The Process

The process of producing leather starts at slaughter. Cows are first stunned, usually with a bolt gun. The gun delivers a blow to the head, making the animal unconscious. Workers hang them by the leg before cutting their throats, causing death by exsanguination. The animal's skin is then removed in one piece and transported to a tannery. Because raw (untreated) hides start to decay within hours, they are layered with salt to draw out moisture and prevent bacterial growth. In some cases, hides are air-dried or sun-dried. There are modern factories that freeze hides to transport them without decomposition. However, salting is the most common method used.

For tanning to begin, workers rehydrate hides by soaking them in large tanks of water. This soaking, taking up to 24 hours, helps to remove dirt, blood and salt residues. They then remove the hair using a **lime** (calcium hydroxide) and **sodium sulphide** bath. The next step is fleshing out a process of scraping excess fat and tissue. At this stage, the hide is called a "pelt."

Lime is very alkaline, having a pH of around 12. This helps to open the chemical structure of the skin. The structure is further broken down with the use of **acidic ammonium salts**, bringing the pelt to a pH of 9-10. Enzymes can then be applied to break down proteins, a step called bateing. The hide is pickled using **sulphuric acid, reducing the** pH to between 3 and 4.

Tanning stabilises the collagen in the hide, preventing it from decomposing. It is a process that requires the soaking of pelts in large baths of chemicals. The time needed for soaking depends on the method and the qualities needed in the leather. The two most common methods are chromium and vegetable tanning. Vegetable tanning, using tannins from tree bark and plants, needs a higher pH and takes weeks to complete. Chrome tanning uses **chromium salts** and is much faster, taking only a few days. But it requires a lower pH of 2-4 and produces toxic waste. Following tanning, the pelt is now technically called leather.

The leather is then dried and split into different thicknesses for various uses. It is the splitting of leather that produces suede and nubuck from the different layers. If the leather needs further processing, it will return to the chemicals to change its properties. At this stage, the leather is then dyed and fat-liquored. Fat-liquoring is applying oils and fats to keep the leather soft and flexible. The leather is then dried again, before buffing, polishing, embossing or adding coatings.

The Environmental Impact

The main waste products resulting from leather processing are salts, alkali (lime), sulphides, acids, chromium salts and dyes. Salts can be fairly easily processed and removed or diluted before release. Acids and alkalis can neutralise each other. However, chromium salts and sulphides are not so easily made safe.

Chromium VI is a carcinogenic byproduct of some tanning processes. It is especially concerning due to its long-term environmental and health impacts. Dyes, finishing agents and waterproofing chemicals can contain hazardous compounds like formaldehyde, arsenic and lead. Without proper treatment, these chemicals leak into surrounding ecosystems, contaminating soil and water supplies.

Resource Use

Water: Leather production is extremely water-intensive, with vast amounts of water used at nearly every stage. In the early phases, water soaks and cleans hides, removes hair and processes raw skins. During tanning, particularly in chrome tanning, large quantities of water are used to create chemical solutions to preserve and treat the hides. Post-tanning processes, such as dyeing and finishing, also demand substantial water use between the chemical treatments before thorough rinsing.

Water Use Comparison: Leather vs. Other Fabrics	
Material	Water Use (Litres per kg)
Leather (chrome-tanned)	15,000 - 50,000
Leather (vegetable-tanned)	10,000 - 30,000
Polyester (synthetic fabric)	50 - 250
Cotton (conventional, non-organic)	7,000 - 29,000
Cotton (organic, rain-fed)	2,000 - 5,000
Wool	10,000 - 100,000+
Linen (Flax)	2,500 - 7,500
Hemp	300 - 2,500
Rayon (Viscose, wood-based fibres)	1,500 - 3,500
Plant-Based Leathers, Mycelium	100 - 1,000

Table 4 [61] [294] [295]

According to the Water Footprint Network, a fully grown beef cow weighing 250 kg will produce 6 kg of leather. This equates to 83% of the water footprint of the animal being attributed to beef. 5.5% of water consumption will be attributable to leather. This results in the water footprint of bovine leather being 17,000 litres per kg. To give this some perspective, an average man's leather jacket will weigh 1.5 kg.

Energy: Leather manufacturing is highly energy-intensive, due to the machinery and heating processes involved. The tanning process uses large industrial drums, mixers and drying systems. All of which consume significant electricity and fuel. Additional energy is used for transportation. Raw hides travel to tanneries and finished leather to manufacturers worldwide. The energy consumption of processing varies depending on the tanning method. For example, chrome tanning, which dominates the industry, is more energy-intensive than traditional vegetable tanning. However, even vegetable tanning has prolonged soaking and drying times, leading to extended energy use.

Much of this energy comes from burning fossil fuels. Coal, oil, or natural gas are burnt to generate heat. This heat helps dry, process, and support chemical reactions in tanning solutions. Additionally, fossil fuels are deeply embedded in the supply chain. They are involved in almost every stage. From petroleum-based synthetic chemicals to the transportation of hides and finished leather products across the globe. Even synthetic coatings and finishes for leather are often derived from petroleum-based sources. These uses show how much the industry depends on fossil fuels.

The energy and fossil fuel use in the manufacture of leather can be compared to those of other fabrics using **Life Cycle Assessments (LCAs)**. These analyse the energy, water and resource consumption at every stage of production.

Below is a general comparison based on available LCA data:

Energy Requirements (MJ per kg of fabric)		
Material	Carbon Footprint (kg CO_2eq/kg)	Estimated Energy Use (MJ/kg)
Leather (chrome-tanned)	17-27	290 – 500+
Cotton (conventional)	5-10	55 – 100
Polyester	2-5	125 – 180
Wool	20-30	150 – 500
Linen (Flax)	1-3	10 – 35
Hemp	2-4	10 – 50
Rayon (Viscose)	5-10	100 – 200
Plant-Based Leather (e.g., Piñatex, Mylo)	1-5	5-50

Table 5

Sources:

M Ulya *et al* , 2021 [300]
Kering Environmental Profit & Loss Report (2018, 2021) [301] [302]
Higg Materials Sustainability Index [303]
Shen et al., 2010 [304]
Laursen et al., 2007 [305]
CarbonFact.com [306]
European Commission, 2023 [307]

Environmental Impact

Water: 10% of leather is produced using vegetable tanning. Almost all the remaining 90% is manufactured using chrome tanning. The wastewater that results is often contaminated with salt, organic waste, heavy metals and other toxic chemicals. Often, particularly in developing countries, this waste is released into local waterways. This severely pollutes water in tannery regions. In countries with weak environmental rules, untreated tannery waste harms local fisheries. It also contaminates soil and causes serious health issues for nearby communities. Workers in tanneries suffer from skin disorders, respiratory illnesses and cancers. They also have a lower life expectancy than those in their communities who do not work in these factories. In Bangladesh, tannery workers rarely live beyond 50 years of age [308].

A 2023 study investigated the levels of heavy metals in the hair of tannery workers in Bangladesh. It found that the workers had "significantly higher levels of chromium in their hair" than people from other countries. The average chromium concentration in Bangladeshi tannery workers' hair was **6.254 mg kg−1**. This is 11.01 times higher than in Poland. It is 20.17 times higher than in Japan. 31.27 times higher than in France and **37.44 times higher than in Sweden** [309]. The presence of chromium at such high levels is indicative of wastewater not being treated or isolated from workers. It is also in Bangladesh that water contaminated with waste chemicals from leather tanning is amongst the highest in the world.

> *"long-term chromium exposure was the primary factor in the majority of the [tannery] workers' high blood pressure, headaches, diabetes, skin allergies, knee pain, jaundice, respiratory issues, and liver abnormalities." – Shimo et al. 2023* [309]

Land: Solid waste from leather production often ends up in landfills. This often includes leftover flesh, trimmings and sludge from wastewater treatment. Sludge from tanneries can contain large amounts of heavy metals. It should be treated and disposed of safely. When this does not happen, for example, it is dumped, it contaminates the soil and leads to heavy contamination of crops. The workers of tanneries then eat these crops, being local to both their work and where food is grown. This only increases their exposure and the levels of heavy metals within their blood and body tissues. As toxic as these metals are for humans, including **Mercury, Arsenic and Lead**, they are just as toxic for plants and animals. Once released into the environment, these dangerous contaminants are not removed from the ecosystem but become integrated. Plants take up the metals. Insects and small animals eat the plants, concentrating the heavy metals in their bodies as they feed. As you look along the food chain, bioaccumulation ensures that the metals become more concentrated in the bodies of predators, poisoning and even killing them. As each organism dies, its body returns to the soil, releasing the metals back into the environment.

Air: Tanneries are usually located far from populated areas. The smells from acids, alkalis, and tanning chemicals make them very unpopular with local communities. Tanneries also cause a lot of air pollution. They release toxic gases, greenhouse gases, VOCs, sulphur compounds, and particulate matter. These pollutants harm human health, damage ecosystems and **contribute to climate change**. Vegetable tanning does produce less air pollution than chrome-tanned leather; however, the processes still emit VOCs and particulate matter.

Volatile Organic Compounds (VOCs) and solvent emissions are released from the chemicals used in the finishing processes of leather. These include **toluene, benzene, xylene** and **formaldehyde**. These chemicals evaporating into the air contribute to **ground-level ozone (smog) and air toxicity**. Prolonged exposure to VOCs causes **respiratory issues, liver damage and increases the risk of cancer**. These compounds also react with nitrogen oxides in the sunlight, forming **photochemical smog**. This smog worsens urban air quality, hovering at low levels where it is impossible to avoid breathing it in.

The dehairing and liming stages of leather production use sodium sulphide and sulfuric acid, which release hydrogen sulphide (H_2S) gas. H_2S is a toxic and foul-smelling gas (like rotten eggs) that can cause eye irritation, headaches, dizziness and even neurological damage at high enough concentrations. When released into the atmosphere, sulphur compounds also contribute to acid rain, damaging soil, water bodies and plant life.

Another foul-smelling chemical, used in the leather tanning process, is ammonia. In tanning, ammonia neutralises acids and adjusts pH levels. Ammonia evaporates easily into the air, especially in warmer climates. Here, it reacts with other pollutants to form **fine particulate matter ($PM_{2.5}$)**. As discussed in Chapter 4, Pollution, Poultry Farming and the Planet, ammonia emissions can lead to the **acidification of ecosystems** and contribute to **eutrophication** (excessive nutrient buildup in water bodies). Additionally, high ammonia exposure can cause **lung irritation** and **exacerbate asthma**.

Tanning leather with chrome also releases **chromium (Cr) vapours** and **formaldehyde** into the air. Chromium, especially in its **hexavalent (Cr^{6+}) form**, is a **known carcinogen**, causing **lung cancer and DNA damage** when inhaled. Formaldehyde, used in **preservatives and coatings**, is also a respiratory irritant and a potential carcinogen.

Grinding, buffing and drying leather produce **particulate matter (PM)**, especially **$PM_{2.5}$** and **PM_{10}**. This fine result is produced regardless of the tanning method. It penetrates deep into the lungs and bloodstream, increasing the risk of **respiratory and cardiovascular diseases**. Also, the dust generated from **chemical residues and dried leather fragments** can be toxic, adding to the health and environmental risks.

Skins

Cows aren't the only animals that are stripped of their skin to make fabric. Other animals, including goats, sheep, horses, pigs, snakes, alligators and deer, are also commodified in this way. Producing skins contributes to **environmental pollution, inefficiency in resource use** and **ethical concerns**. While some of these skins are by-products of the meat industry, many are produced **solely for fashion**, making their production even more wasteful.

Snakeskin (Pythons, Cobras and **Other Reptiles):** Snakeskin is made from either wild or farmed snakes. Both practices are harmful to the reptiles. Wild snake populations are declining due to climate change, hunting and other human activities, and farmed snakes have high mortality rates, caused by poor living conditions. Processing snake skins requires heavy chemical processing due to the delicate nature of snake and reptile skins. This can involve the use of **arsenic-based preservatives**, which pollute water and air. The snake and reptile skin market is also highly dependent on the exporting of products, leading to **high transport emissions**. Using snakes and other reptilian animals is extremely inefficient, as their bodies provide **very little usable leather**. A single python may only produce enough skin to make a small handbag or a few belts, making it highly **resource-inefficient**.

Goat Leather (Kidskin): Goats need a lot of water, especially in dry countries where they are often farmed. Also, like the farming of sheep, goat farming **degrades soil** due to their feeding habits. This is leading to **desertification** in places like **India, Pakistan and parts of Africa**. Kidskin (the skin of baby goats) is often **chrome-tanned**, contributing to **heavy metal pollution**. Goats are also fairly small animals. This means that many goats will need to be killed to produce enough fabric to make items, increasing the inefficiency of the process.

The Pollution and Inefficiency of Animal Leathers

Animal	Pollution Level	Inefficiency
Alligator & Crocodile	🔴 **Extremely High** (chemical-heavy, water-intensive, wildlife destruction)	🔴 **Very Inefficient** (slow-growing, small yield)
Snakes & Reptiles	🔴 **Very High** (arsenic tanning, habitat destruction)	🔴 **Highly Inefficient** (tiny leather yield)
Pigskin	🔴 **High** (ammonia, manure pollution, chemicals)	🔴 **Inefficient** (better plant-based alternatives exist)
Sheepskin	🟠 **High** (methane emissions, acid tanning)	🟠 **Moderate Inefficiency** (soft but high-maintenance)
Goat Leather	🟠 **Moderate to High** (overgrazing, chemical waste)	🟠 **Moderate Inefficiency** (small skins, high processing needs)
Horse Leather (Cordovan)	🟡 **Moderate** (chemical-heavy processing)	🔴 **Highly Inefficient** (limited skin area used)

Table 6

Sheep Leather (Sheepskin & Suede): Sheep are ruminants. They are major contributors to **methane pollution** (second only to cows), making them **high-emission livestock**. Also, the **soft nature of**

333

sheepskin requires **strong chemical treatments** to strengthen it. This results in **acidic wastewater and VOC emissions**. Many of the sheep killed for their skin are part of the wool industry. We've already been over this very chemical-dependent and resource-intensive sector. Producing sheep skin is another inefficient way of manufacturing fabric. Using land, feed, pesticides and water inputs to grow an entire animal to make use of their skin is wasteful. Even if the sheep meat industry were to decline, a market for sheep skin and wool-based yarns would still exist. The practice of breeding and raising sheep as a product would continue.

Pig Leather (Pigskin): We explored the damage caused by pig farming in Chapter 7, Hogs and Harm. How resource-intensive it is to produce pigs as a commodity. This included the use of water, land and feed. Pig farming releases **ammonia**, **hydrogen sulphide** and **methane**, causing **air** and **water contamination**. The large amounts of manure that pigs produce cause the contamination of soil and water. Also, pigskins need **heavy chemical processing**, releasing **chromium, formaldehyde** and **synthetic dyes** into the environment.

Horse Leather (Cordovan): Cordovan leather is **highly processed**. Rather than weeks, it needs **months of tanning** and many chemical treatments. Also, as cordovan comes from only a specific part of the horse's hindquarters, **most of the hide is wasted**.

Alligator & Crocodile Leather: Most alligator and crocodile leather comes from farmed animals. But **illegal hunting** still threatens wild alligators and crocodiles. Alligator farms, especially in **Southeast Asia** and **Florida and Louisiana, USA**, need **massive amounts of water** and **feed**. Crocodile leather is **thick** and **heavily scaled**. This means it requires harsh **sulphuric acid and chrome tanning** to make it usable. Also, alligators grow **slowly**. They eat **large amounts of food** while **producing relatively little leather**. This makes them extremely **resource-intensive**.

Sustainable Alternatives

With growing awareness of the environmental impact of leather, many companies are developing sustainable alternatives. Options like Piñatex (pineapple leather), apple leather and mushroom leather use agricultural byproducts. They reduce waste while avoiding the environmental harm of animal agriculture. For example, Piñatex, made from the fibres of pineapple leaves, is a by-product of the fruit industry. Apple leather is derived from apple pomace left over from cider production.

> *In Bangladesh, 85,000 tons of rawhides and skins are processed annually. Processing one metric ton of rawhides produces 200 kg of leather, 250 kg of non-tanned solid waste, 200 kg of tanned waste, and 50,000 kg of wastewater. Together these create 8 kg of chromium. – Hira et al. 2022* [310]

Cactus and cork leathers are also popular, requiring minimal resources to produce. Additionally, recycled and bio-based synthetic leathers provide durable alternatives without relying on fossil fuels. These materials eliminate the pollution, deforestation and water consumption linked to animal leather while preventing the ethical concerns of factory farming.

Animal Leathers and Plant-Based Alternatives - A Side-By-Side Comparison		
Factor	Animal Leathers (Cow, Sheep, Pig, Reptile, etc.)	Plant-Based Leathers (Cactus, Mushroom, Pineapple, etc.)
Raw Material	Animals (cows, pigs, goats, snakes, alligators, etc.)	Plants (cactus, mushrooms, pineapple leaves, apple waste, etc.)
Land Use	**Very high** (Deforestation for grazing, farmland needed)	**Low** (Uses agricultural waste or fast-growing plants)
Water Use	**Extremely high** (Cattle farming and tanning require vast amounts of water)	**Low** (Requires little irrigation, often rain-fed)
Fossil Fuel Use	**High** (Livestock feed production, processing, transport)	**Low to moderate** (Some require minor synthetic binders)
Greenhouse Gas Emissions	**Very high** (Methane from livestock, CO_2 from tanning, transport emissions)	**Very low** (Minimal emissions, some CO_2 absorption during plant growth)
Chemical Use	**Toxic chemicals** (Chrome, formaldehyde, sulphur compounds, ammonia)	Non-toxic or biodegradable chemicals (Corn-based binders, natural dyes)
Air Pollution	**High** (VOCs, hydrogen sulphide, methane, dust particles)	**Low** (Some VOCs from certain processing methods)
Water Pollution	**Severe** (Heavy metal contamination, acid runoff, manure waste)	**Minimal** (Mostly biodegradable residues)

Table 7

Leather made from animal skins cannot be directly compared to plant-based leather alternatives. This is because the raw materials and processing techniques vary considerably from one product to another. Basic comparisons are possible, using the metrics of input versus output and how much waste results per product:

A **cost comparison** between **animal leather** and **plant-based alternatives** also shows how harmful animal-based leathers are.

Cost Comparison: Animal Leather vs. Plant-Based Alternatives		
Factor	Animal Leather (Cow, Sheep, Pig, Exotic)	Plant-Based Leather (Cactus, Mushroom, Pineapple, etc.)
Raw Material Cost	**High** (Feed, water, land, labour)	**Low to moderate** (Uses agricultural waste or fast-growing plants)
Energy Cost	**Very high** (Farming, slaughter, tanning require large energy inputs)	**Low to moderate** (Little energy required, mycelium needs controlled environments)
Water Cost	**Extremely high** (Farming and tanning)	**Low** (Minimal irrigation needed, often rain-fed)
Chemical Cost	**High** (Chromium, formaldehyde, acids, dyes)	**Low** Many are non-toxic
Waste Management Cost	**Very high** (Toxic sludge disposal, regulations, waste treatment)	**Low** (Mostly biodegradable and non-toxic waste)
Retail Price per Square Metre	$30-$100 (Standard cowhide), $200-$500 (Exotic leathers like alligator)	$25-$70 (Cactus, mushroom, or pineapple leather)
Environmental Cost	**Extremely high** (Deforestation, methane, toxic waste, water pollution)	**Low to moderate** (Some synthetic elements, but far less damaging)

Table 8

Fur: The Environmental and Ethical Issues

Around 100 million animals are killed each year for the global fur trade, with 15% trapped in the wild and the remaining 85% reared on fur farms. Fur has long been a symbol of luxury and status, but its production is fraught with both environmental and ethical issues. This has led to the fur industry facing increasing scrutiny in recent years, due to concerns about animal welfare, habitat destruction and pollution. In recent times, fur has seen a decrease in popularity, as the public has become aware of the inherent cruelty involved in its production. The market for fur further declined in 2020 after outbreaks of SARS-CoV-2, the virus that causes COVID-19, on 400 mink farms in Europe and America. The infected animals led to more than 200 cases of human infection, all of which were directly linked to fur farms. These infections culminated in the culling of over **20 million animals** [311].

The History of Fur Use

Humans using fur dates back tens of thousands of years. It played a crucial role in survival, culture and later, fashion. Early humans relied on fur for warmth in colder climates, using the pelts of animals they had hunted for food. There is evidence that during the Palaeolithic era (roughly 2.58 million to 11,700 years ago), people were making clothing from animal hides. They invented tools to scrape and soften them. Over time, techniques improved, and fur became a necessity for some, but also a symbol of status and identity.

In ancient civilisations, fur was associated with power and wealth. Egyptian, Greek and Roman societies valued fur for its warmth and luxury, while reserving the finest pelts for royalty and nobility. During the Middle Ages and the Renaissance in Europe, the fur trade flourished. Beaver, sable and ermine pelts became highly sought after. The demand for fur led to extensive trapping and hunting, shaping economies. The trade even fuelled exploration, as European traders travelled to North America and Siberia in search of valuable furs.

By the 18th and 19th centuries, the fur trade had become a global industry. In North America, it contributed to the exploitation of Indigenous peoples. They were often forced into fur trading relationships that disrupted their traditional ways of life. As industrialisation progressed, fur remained a symbol of status. Mink, fox and chinchilla coats were in high demand among the wealthy.

In the 20th century, fur farming expanded, replacing traditional hunting with large-scale breeding. But growing awareness of animal cruelty and concerns about environmental damage created opposition to the industry. By the late 20th and early 21st centuries, animal rights activists began challenging the fur industry. Their work included protesting outside shops selling fur. They would throw red paint over people wearing fur outside, including celebrities. This made people fearful of being seen wearing the fabric. These actions led to stricter regulations, bans and a steep decline in popularity.

The biological functioning of mink and foxes farmed for fur is impaired, as indicated by levels of stereotypic (abnormal repetitive) behaviour, fur-chewing and tail-biting/self-injury, physical deformities (bent feet) and high levels of reproductive failure/infant mortality; - Pickett & Harris, 2015 (312)

Pelt Production

Fur farming involves breeding and raising animals specifically for their pelts, with the most farmed species being mink, foxes, chinchillas and raccoon dogs. These animals are kept in barren wire mesh cages, often stacked in rows inside large sheds or outdoors in open-air enclosures. The conditions are typically cramped, with animals allowed only enough space to stand up and turn around, preventing them from engaging in natural behaviours such as hunting, digging, or swimming. For example, cages are typically 90 cm deep and 30 cm wide for mink, allowing the animals to move no more than four paces before reaching the end of the cage. In the wild, mink will normally travel between 1.1 and 7.5 km from their den. Cages for foxes are usually 0.8-1.2 m², not in any way close to the more than 5 km a wild fox would travel in a day [312].

As fur quality is the priority, selective breeding is used to produce animals with desirable coat colours and textures, leading to genetic issues like weakened immune systems and deformities. Arctic foxes are bred to grow excessive amounts of skin to increase the size of the pelt. This results in skin folds that can lead to skin diseases and often cause skin to fold over their eyes, making fur rub against their eyeballs. Breeding females are often kept in isolation except during mating, and their offspring are raised in captivity until they reach the ideal age for slaughter, usually a few months old.

The killing process on fur farms is designed to minimise damage to the pelts, with little to no concern for animal welfare. Common methods include gassing (used primarily for mink), anal or genital electrocution (frequent for foxes and chinchillas), or breaking the animals' necks. These methods are meant to preserve the fur but often cause immense suffering. Some animals are skinned alive, as this is falsely believed to improve the quality of the pelt. Once the animals are killed, their bodies are discarded, either by burning or being turned into a variety of products from components of tyres, to feed for zoo-kept or farmed animals. The pelts will undergo a tanning and chemical treatment process like that of leather to prevent decay and make the fur more durable.

In addition to fur farming, wild animals are still trapped for their pelts, with species like bobcats, beavers, lynxes and martens being among those targeted. Trapping methods include steel-jaw leghold traps, body-gripping traps, drowning traps and snares, all of which cause severe pain and suffering. Many trapped animals experience broken bones, blood loss or starvation after being held in traps for many hours or even days, before trappers return to kill them, often by blunt force trauma or suffocation. Non-target animals, including endangered species and companion animals, frequently become unintended victims of these traps. Despite the cruelty involved, wild-caught fur remains highly valued in the fashion industry due to its perceived authenticity.

The Process

Processing fur into fabric involves several steps. Each one works to preserve the pelts, enhance their durability, and prepare them for use in clothing and accessories. The steps include skinning, curing, dressing, dyeing and sewing the furs together into usable fabric.

Skinning and Curing: After slaughter, the pelt of an animal is carefully removed, avoiding damaging the fur. The skin is then stretched, cleaned and salted to prevent decomposition.

Dressing and Tanning: Once cured, the pelts go through a chemical treatment process called "dressing". This makes the skin soft, durable and resistant to decay. Hides are soaked in solutions that remove any remaining flesh, fat or unwanted debris. They are then tanned, using **formaldehyde, chromium salts**, or vegetable-based tannins. These prevent rotting and maintain flexibility. These chemicals can be highly toxic, contributing to environmental pollution. Following tanning, the skins go through stretching and softening, changing their texture.

Dyeing and Finishing: Many furs keep their natural colours. But some are dyed to match fashion trends or create uniformity in the final product. This creates one colour in a fabric made from the many skins sewn together. This process creates larger pieces of fabric from small animal skins. The dyeing process uses chemicals that penetrate the skin and fur without changing the texture. After dyeing, treatments, such as waterproofing or anti-shedding sprays, improve the fur's appearance and durability. All of which prevent fur from being biodegradable.

The Environmental Impact

There are no positive impacts from fur farming. The negative environmental impacts of the fur industry are like those of the leather industry. This overlap comes from the tanning to preserve skin. The process uses large amounts of salt to cure pelts. This is often washed into waterways during the removal stages. Chemicals used include formaldehyde, **aluminium sulphate**, **lime**, **sulphuric acid** and **chromium salts**. **Synthetic dyes and pigments**, such as **naphthalene**, a known **carcinogen**, colour fabrics for clothing and other items.

The farming of animals of any kind results in organic waste, manure and carcasses. When people hunt wild animals for fur, their removal can have devastating effects on ecosystems. This results from disruption to food chains and other natural processes. Trapping reduces the populations of targeted species, causing harm to non-target animals. This includes endangered animals when they get caught in traps. There are smaller changes, such as the stool of one animal fertilising the plant that is food for another. In regions where coyotes are the target, grey wolves also become ensnared in leghold or body traps. These endangered wolves have decreased in number;

some countries have fewer than 100 wolves left. Others have none; they have wiped them out

Ammonia Pollution from Fur Farming

In 1996, Bäck et al. published a paper investigating the impact of fur farms on pine trees in the Netherlands [313]. The study found that pine forests suffered from overexposure to nitrogen. This caused the trees to suffer from nutrient imbalances that damaged the health of their foliage. The result was the same whether the nitrogen came from overuse of fertiliser or natural accumulation. This can lead to decreased survival when trees suffer increased stress. This included episodes of frost or drought. The study found trees developed unhealthy shoots even when the weather improved. High levels of nitrogen also result in microflora, such as algae, on the surface of leaves. These tiny plants covering leaves limit their ability to absorb and emit gases and water vapour. Combined, these changes reduce photosynthesis, limiting tree growth and survival.

When examining land close to large fur farms in Finland, they found many areas of forest had become grassland. Trees couldn't survive in the acidified soils resulting from ammonia releases. The remaining trees showed damage caused by nitrogen overload and ammonia toxicity. They also had a build-up of protein within their leaves. Close examination of the leaves found that trees growing near fur farms had changes in their needle anatomy. This included thinner cytoplasm in mesophyll cells. They saw folded plasmalemma and undulating chloroplast membranes. These changes indicated that trees were suffering from stress. The stress was so great that it interfered with normal processes like photosynthesis and nutrient transport.

The paper underscores that fur farming, through ammonia emissions, contributes to environmental pollution. These releases extend over large distances, affecting the health of ecosystems, including forests.

Alternatives to Fur

Animal fur has already started to fall out of fashion. Since the 1980s, there has been a steady decline in the number of people wanting to wear fur, due to animal welfare concerns. Globally, in 2012, the production of mink and fox pelts exceeded 81 million. This had decreased to fewer than 15 million in 2023, as many countries ended production. The Netherlands ended fur farming in 2021, following the mass culling of mink due to infections with COVID-19. Norway and Finland both ended fur farming in 2018, Austria in 2005 and the UK in 2003. Romania, Latvia and Lithuania have already legislated to ban fur farming in 2027 and 2028 [314]. However, we need a global ban on commercial fur production for the sake of the planet and animals.

One argument against fur alternatives is that they are not organic. They do not break down or rot. But neither does tanned fur, at least not as fast as untreated fur would. The chemical processes involved in tanning pelts preserve the skin. They permanently change the bonds in the skin, preventing decomposition. When discarded, fur will remain for years, slowly breaking down, contaminating the ground when sent to landfill. British Fur, the trade association, states that fur "can last a lifetime". There has been some effort to recycle fur products. But this often involves more chemicals to restore the look and feel of fabrics to something desirable.

Alternatives to real fur have been available for decades. Made from plastic-based synthetic fibres, these fabrics are a popular alternative to real fur. They can replicate the look and feel of the fabric without the associated ethical and environmental issues. But, made from fossil fuels, synthetic fur products are not environmentally neutral. These products, though, are still less harmful to the environment than animal-based products.

It's possible to compare methods of fur production and the resulting environmental impacts. The table below gives examples using the metrics of fossil fuel and water use and air, land and water pollution:

Impact	Mink Fur	Fox Fur	Rabbit Fur	Faux Fur (Virgin Synthetics)
Fossil Fuel Use	High – Intensive for farming and feed production	High – Similar to mink, with substantial energy use	Moderate – Smaller animals need less feed, farming still needs fossil fuels	High – Energy intensive from synthetic fibres production
Water Use	Very High – Farming and processing require substantial water	High – Similar to mink, with significant water use	Moderate – Considerable due to farming needs	Moderate – Water used in fibre production and dyeing processes
Air Pollution	High – Emissions from manure and chemical processing	High – Similar to mink, with emissions from waste and chemical processing	Moderate – Smaller scale operations result in lower emissions	Moderate – Manufacturing synthetic fibres releases pollutants
Land Pollution	High – Waste from farms, and chemicals	High – Similar to mink, with waste and chemical runoff	Moderate – Less waste, but farming practices still impact soil health	Moderate – Microplastics from synthetic fibres can persist in soil
Water Pollution	Very High – Farm runoff and chemicals severely pollute water	High – Similar to mink, with water contamination from waste and chemicals	Moderate – Water pollution from farming runoff and chemicals	High – Production processes can release microplastics and chemicals

Table 9

Sources:

PETA [315]

Fur Free Alliance [316]

Ecopel [317]

HuffPost [318]

Collective Fashion Justice [319]

As the table shows, there is no "solution" to the problem of people wanting to wear fur. All production methods are polluting and resource-intensive. It is possible to produce faux fur from recycled plastics. But this is an emerging industry, only accounting for a small percentage of fur-type fabrics produced. Using recycled plastics as the main component of faux fur is better for the environment. It uses less fossil fuels and water while resulting in less pollution. But right now, these products are not readily available. This means that the most environmentally friendly way to wear fur is to continue using the fur items that you have. This applies whether they are faux fur or animal skin. You can donate what you no longer want to charity, swap with friends or sell items and buy second-hand.

Feathers

Humans Using Feathers

Humans have been using bird feathers for thousands of years. From decoration and symbolism to practical uses, feathers are culturally significant.

People have used feathers as decorations in clothing, headdresses, and jewellery for thousands of years. Indigenous cultures worldwide, including Native American, Aztec and African societies, have worn feathers since ancient times. They were used as symbols to signify status, bravery, or spiritual connection. Many cultures also viewed feathers as sacred, believing they carried spiritual significance. For example, Ancient Egyptians associated feathers with Ma'at, the goddess of truth and justice. In Christianity, feathers have a connection with angels.

In Europe, in the 6th century, people made quill pens with feathers. This was until metal dip pens replaced them, much later, in the 19th century. In the medieval and Renaissance eras, feathers decorated the hats, capes and clothing of nobility. In the 19th and early 20th centuries, the demand for feathers peaked in the Victorian and Edwardian eras. During this time, women's hats featured elaborate feather decorations. This led to the mass hunting of birds, causing the decline of species like the snowy egret.

Since this period, feathers have been used for practical purposes. They add warmth to items like down bedding and outdoor clothing. Fishers practising fly fishing still rely on intricately tied feather lures to mimic insects.

The Process

The commercial feather industry primarily sources feathers from domesticated birds. These include chickens, ducks, geese and turkeys. The most common feather used is the down feather. These are the soft, fluffy feathers that provide birds with insulation. Larger flight feathers are also used for more aesthetic purposes. Feathers are a byproduct of the poultry industry. Feathers collected this way come from birds slaughtered for meat. However, in some cases - particularly in the down industry - live-plucking still happens. In this process, birds, especially geese, have their feathers pulled out while still alive. This allows workers to collect their down several times before slaughter. This is an extremely controversial practice. Due to animal welfare concerns, some producers will lie about the origins of the feathers. They will state that they only use feathers from birds that were not live-plucked. They do this knowing that, if buyers knew the truth, the feathers would not sell.

Once collected, feathers undergo thorough cleaning to remove dirt, blood, and other contaminants. This involves washing them with detergents and disinfectants, removing bacteria and parasites. The feathers then dry at temperatures of up to **120°C**. This further sterilises them and prevents mould or bacterial growth. After drying, they are sorted by size, type and quality. Down feathers, desired for their insulation properties, are separated from stiffer flight feathers using air-blowing machines or mechanical sorting methods. High-quality down is often graded based on its loft (the ability to trap air and provide warmth), with the finest down sourced from mature birds.

Various industries use processed feathers. Down feathers become filling for bedding, jackets and sleeping bags. Stiffer feathers make fashion accessories, decorative arts, or practical items like fishing lures or arrow fletching. Some industries also use ground-up feathers as an ingredient in animal feed or as a source of keratin in certain industrial products.

Live Plucking

We have already explored the environmental impacts of farming poultry. They include breeding, housing, feeding, and ultimately killing the birds. These consequences also apply to the feather market. Keeping poultry animals uses vast quantities of resources. It results in air, water and soil pollution, as well as direct harm to the workers. These people suffer from respiratory diseases, abnormal inflammatory responses and bacterial infections. This harms both staff and the lives of those living near poultry farms. In the feather market, the belief by many is that feathers come from birds slaughtered for meat. The feathers are a by-product. This does happen, to some degree. But, more than 90% of feathers, used for clothing, bedding and decorative purposes, come from China. In this country, live plucking is a common industry practice.

Live plucking is exactly what it sounds like. It is a process of ripping feathers from live birds, mostly geese and ducks, while they remain fully conscious. Workers use the birds as feather-producing factories. After they pull the feathers out, they leave the birds to regrow them before ripping them from the birds' bodies again. They repeat this several times throughout the bird's life. It causes extreme distress, open wounds and long-term suffering. During feather removal, a manual process, people restrain birds with a foot over their necks. Feathers are then ripped out by hand, sometimes causing their skin to tear. Afterwards, injuries are hastily stitched up by workers with no veterinary training and no suitable equipment. They just use a regular needle and thread. They do this with no pain relief for the bird before returning them to the flock to heal enough to repeat the process.

Many birds in the feather market, particularly ducks and geese, are also force-fed as part of the **foie gras** industry. This practice, gavage, which means force-feeding, involves workers inserting a metal tube down the bird's throat to pump large quantities of grain and fat into their stomachs. The birds are put through this process many times a day. The overfeeding enlarges their livers, which are then harvested for foie gras. Birds raised for foie gras are often used in the feather industry as the process allows for repeated feather plucking while birds are alive. This maximises profits from each bird before they are slaughtered for their livers.

Despite international pressure and claims of ethical sourcing by many brands, investigations have repeatedly exposed that live plucking is still widely used. China is the largest supplier of down feathers. Some Chinese producers claim to source only from slaughtered birds (a practice known as "post-mortem plucking"). But undercover investigations have shown that live plucking is still widespread. The sheer volume of feathers needed to meet global demand means that many suppliers resort to live plucking. They prefer this method as it allows them to pluck birds multiple times rather than relying on a single harvest from dead birds.

Many companies have also been exposed for misleading consumers about their sourcing practices. After stating they only use feathers from slaughtered birds, they have been caught in a lie. Reports and undercover videos show that big brands use farms that practise live plucking. These are brands claiming their down is "ethically sourced". Certification programs such as the Responsible Down Standard (RDS) and the Global Traceable Down Standard (TDS) have attempted to reassure consumers. But they have faced criticism for failing to prevent live-plucked feathers from entering supply chains. This is a hard task as these feathers are not distinguishable from post-mortem plucked feathers. Loopholes in auditing let dishonest suppliers mix feathers. This makes it tough for companies to check where their materials come from. Many consumers buy products with feathers without knowing the pain involved. Meanwhile, companies keep profiting from misleading marketing.

Pain-Free Alternatives

The overuse of feathers, especially in fashion, led to conservation efforts in the early 20th century. The Migratory Bird Treaty Act of 1918 in the U.S. made it illegal to hunt certain birds for their feathers. Today, ethical concerns have increased the use of synthetic feathers in fashion and bedding.

The obvious alternatives to feathers, including insulation, are synthetic fibres such as polyester. These fibres, made from fossil fuels, are not sustainable. Yet they are still less polluting than farming birds and obviously less cruel. There are also sustainable plant-based alternatives to feathers available. They use less water while having similar properties to down feathers. Using warmth and water resistance as benchmarks, feathers are warm but not very water-resistant. Kapok fibre, one natural alternative, is not quite as warm as feathers, but it is more water-resistant. Then there is milkweed fibre, which is as warm as feathers and more resistant to water.

Material	Warmth	Breathability	Water Resistance	Sustainability	Water Use
Feathers (Down)	●●●●●	●●●●●	●●●	●	●●●●●
Kapok Fibre	●●●●	●●●●●	●●●●	●●●●●	●
Hemp Fibre	●●●	●●●●●	●●●	●●●●●	●●
Flax Fibre	●●●	●●●●●	●●	●●●●●	●●●
Organic Cotton Batting	●●●	●●●●	●●	●●●●	●●●●
Banana Fibre	●●●●	●●●●	●●●	●●●●●	●●
Coconut Fibre (Coir)	●●	●●●	●●●●●	●●●●●	●
Milkweed Fibre	●●●●●	●●●●	●●●●●	●●●●●	●
Algae-Based Insulation	●●●●●	●●●●●	●●●●●	●●●●●	●

Table 10

NASA

Those who study thermodynamics know that feathers have limited thermal properties. This is why anyone who needs reliable thermal protection, such as NASA, the space agency, does not use feathers to insulate any of their equipment [320]. NASA doesn't use feathers for insulation in space for several key reasons:

1. Lack of Extreme Insulation Performance: Feathers **don't perform well in the vacuum of space**. This is because heat transfer occurs through radiation rather than conduction or convection.

2. Moisture Absorption and Degradation: Feathers **absorb moisture**. This can lead to **mould, bacterial growth and reduced insulation effectiveness**. This would be a major issue in space, where maintaining a dry environment is crucial.

3. Weight and Bulk: Space missions prioritise lightweight, compact materials. Feathers, though light, **take up more space** compared to high-performance insulations like aerogels. These other materials also provide better insulation at a fraction of the thickness.

4. Durability and Longevity: Feathers **break down over time.** NASA needs highly durable materials. They must also be non-degrading and withstand years of space travel and extreme temperature shifts.

5. Fire Safety Risks: Feathers are **flammable**. They would be a fire is a huge risk in enclosed space environments like the ISS. NASA only uses materials that meet strict **non-flammable safety standards**.

6. Limited Customisation: NASA engineers insulation materials to **precisely control heat transfer**. Feathers provide only **passive insulation** with less flexibility in performance tuning.

Our daily lives do not need fibres that meet the standards needed in space. But it does seem that people use feathers by the masses for convenience. Birds produce feathers that are useful, and so we take them for our purposes. We literally rip them from the birds' backs and stomachs.

Conclusion

It is an outdated idea that we need to use animals for fabric. It is unnecessary and deeply harmful. Today, we have access to a wealth of plant-based and innovative materials. These fibres provide warmth, durability and comfort without the cruelty or destruction associated with wool, silk, leather and down.

Fibres from plants like hemp, flax, organic cotton and banana leaves prove that we don't need to exploit animals for clothing. Emerging materials like algae-based insulation and lab-grown leather provide even more options. These alternatives are not just ethical - they are also more sustainable.

The environmental impact of animal-derived fabrics is staggering. **Wool** production leads to massive **land degradation, deforestation** and **water pollution**. Farming sheep releases **methane**, a potent greenhouse gas. **Leather** production causes **deforestation**, particularly in the Amazon, where farmers clear land to graze cattle and grow feed. The tanning process involves **toxic chemicals** like chromium. These poison waterways and harm workers in the industry. **Silk** production **boils billions of silkworms alive**. Down **feathers** often come from birds that have endured the excruciating pain of **live plucking**. Every step of these industries exploits animals while damaging the planet.

Choosing plant-based materials ends the unnecessary suffering and reduces our environmental footprint. The technology already exists to replace every animal-derived fabric with sustainable, cruelty-free alternatives. What's missing is the recognition that animal fabrics are neither essential nor justifiable. By eliminating the use of animal-derived fabrics, we protect ecosystems. We reduce pollution and spare countless animals from unnecessary suffering. Sustainable alternatives like plant-based leathers, recycled synthetics and pioneering fibres are cruelty-free. They are environmentally friendly solutions, and they align with a future where fashion and textiles do not come at the cost of animals' lives, human health, or the planet.

Chapter 10: It's Not Just Animal Agriculture That's The Problem

Animal agriculture produces more emissions than all methods of transportation combined. This is a phrase often used in vegan discussions. Those who take an anti-vegan stance argue that this is inaccurate at best, ridiculous at worst. Dan Blaustein-Rejto wrote an article L*ivestock Don't Contribute 14.5% of Global Greenhouse Gas Emissions*. He also wrote *Cutting Beef Isn't the Only Way*. In the emissions piece, he said the figures given for emissions resulting from animal agriculture cannot be trusted. That different sources produce different numbers [321]. Rejto says the Food and Agriculture

Organization of the United Nations estimates that livestock contribute 11.1% to global greenhouse gas emissions. But peer-reviewed studies found that the number was higher at up to 19.6% of the total. The article does accept that the demand for meat could be met with plant-based alternatives. He also agreed that methane and nitrous oxide emissions from animal agriculture are a significant problem. But the article says that methane emissions *"do not need to be eliminated"*. Instead, cows should eat seaweed additives to reduce greenhouse gas releases from digestion. He reasons that people rely on farming animals for their livelihoods. Rejto does not consider a shift away from animal farming to growing crops for humans.

Whether animal agriculture contributes 11.1% or 19.6% to the total of global greenhouse gas emissions each year, there is still more than 80% of emissions from other industries unaccounted for. The total global greenhouse gas (GHG) emissions can be broken down by sector as follows:

1. **Energy (73.2%)** - The largest source of global emissions, mainly from burning fossil fuels for electricity, heat and transport. This includes:
   - **Electricity & Heat Production (30.4%)** - Power plants burning coal, oil and gas.
   - **Industry (24.2%)** - Emissions from manufacturing, cement, steel and chemical production.
   - **Transport (16.2%)** - Road, air and sea transport, primarily fuelled by oil.
   - **Buildings (17.5%)** - Energy use in residential and commercial buildings for heating, cooling and lighting.

2. **Agriculture, Forestry and Land Use (18.4%)** - Includes deforestation, livestock farming (methane from cows and manure), rice production and soil emissions from fertilisers.

3. **Industry (5.2%)** - Non-energy-related emissions from processes such as cement production, which release CO_2 during chemical reactions.

4. **Waste (3.2%)** - Emissions from landfills (methane), wastewater treatment and the burning of waste materials.

Sources:

Our World in Data [322]
IPCC Reports [323]
Climate Watch (World Resources Institute - WRI) [324]
International Energy Agency (IEA) [325]
FAO (Food and Agriculture Organization) [326]

Energy

"almost three-quarters of emissions come from energy use; almost one-fifth from agriculture and land use [this increases to one-quarter when we consider the food system as a whole - including processing, packaging, transport and retail]; and the remaining 8% from industry and waste." – Hannah Ritchie [327]

In 2020, Hannah Ritchie wrote the article *Sector by sector: where do global greenhouse gas emissions come from?* She uses the 2016 data from Climate Watch to examine greenhouse gas emissions from different industries [327]. At the time of writing, this data is 9 years old. But, since 2016, the share of greenhouse gas releases has not changed much. The latest data from the European Commission's **Emissions Database for Global Atmospheric Research (EDGAR)** states that the total global greenhouse gas emissions for **2023** were **52962.90** Mton CO_2eq. The total greenhouse gas emissions in **2015** were **48808.77** Mton CO_2eq. This is an increase of 7.85% in 8 years. Yet, how the emissions divide between sectors has not changed significantly [328].

Out of the total greenhouse gas emissions, roughly 73% come from energy production. Just over 24% comes from industry. 17.5% results from energy used to power buildings, and a little over 16% comes from transportation. This includes road, aviation and shipping sectors.

Industry Energy Use

The industrial sector relies on energy for various processes. They include manufacturing, mining, refining, and construction. This energy powers machinery, heats materials, drives chemical reactions, and operates industrial facilities. The primary energy sources in industry include fossil fuels (coal, oil, and natural gas). Electricity is also used and, to a lesser extent, renewable energy sources. The choice of energy source depends on the specific requirements of the industrial processes. High-temperature processes often rely on fossil fuels due to their ability to supply energy at a consistent rate and high density.

A large part of industrial energy use comes from the steel, cement and chemical production sectors. Steel manufacturing relies on blast furnaces fuelled by coal and natural gas, releasing large amounts of CO_2. Similarly, cement production is highly energy-intensive. It requires high temperatures for the chemical conversion of limestone into clinker. This process, on its own, contributes roughly **8%** to global CO_2 emissions. The chemical industry also consumes vast amounts of energy. The production of ammonia, plastics and fertilisers needs both heat and hydrogen derived from natural gas.

Efforts to decarbonise industrial energy use focus on improving energy efficiency. Companies are switching to low-carbon fuels and electrifying processes where possible. Additionally, industries are investing in renewable energy sources, such as green hydrogen, to replace fossil fuels.

Energy Use in Buildings

Energy use in buildings accounts for approximately **6.6%** of global greenhouse gas emissions coming from fuel combustion. This share increases when considering indirect emissions from electricity and heat production. Buildings need energy for heating, cooling, lighting, cooking, water heating, and operating appliances. The energy sources used vary by region. Some countries use greater amounts of renewable energy, including wind and solar. But in many countries, fossil fuels are still the main energy source for buildings, particularly for space and water heating.

Residential buildings tend to use more energy for heating and cooling due to the inefficiency of heating many smaller units. Commercial buildings - such as offices, schools, hospitals and retail spaces - consume more energy for lighting, ventilation and electronic equipment. Many factors influence the demand for energy in buildings. These include the climate, insulation quality, construction materials and the efficiency of heating, ventilation and air conditioning (HVAC) systems. Poorly insulated buildings use more energy for heating in winter and cooling in summer, increasing energy consumption and emissions. Energy-efficient buildings with proper insulation, passive design strategies and advanced HVAC systems can use far less energy.

The process of decarbonising building energy use is underway across the world. Key strategies include improving energy efficiency, transitioning to renewable energy and adopting smart technologies. Energy efficiency measures, like LED lighting and high-efficiency appliances, can reduce energy demand. Using heat pumps for heating and cooling instead of gas furnaces can reduce emissions. This improves further when powered by renewable electricity. Additionally, on-site renewable energy generation, such as rooftop solar panels, can help buildings become more self-sufficient and reduce reliance on fossil fuel-based energy grids.

Energy Use in Transportation

The transportation sector is a major consumer of energy, producing roughly **16%** of global greenhouse gas emissions. The movement of transport, including cars, planes, and ships, relies on petroleum-based fuels, such as petrol, diesel, and jet fuel. Road transport is the largest contributor. It produces nearly three-quarters of transport emissions, with passenger vehicles and freight trucks being the largest sources. Air travel and shipping also contribute substantially, as they rely on high-energy-density fuels that are difficult to replace with low-carbon alternatives.

Factors such as vehicle efficiency, infrastructure, and travel demand drive energy use in transportation. Internal combustion engine (ICE) vehicles are inherently inefficient. Most of the fuel's energy is lost as heat rather than producing motion. Efforts to reduce transportation emissions focus on improving fuel efficiency. Many countries are expanding public transit, especially trains. They are also shifting to low-carbon fuels like biofuels, hydrogen, and electricity.

The Role of Electric Vehicles

Electric vehicles (EVs) are a key technology in reducing transportation emissions. These vehicles can replace petrol and diesel-powered cars with battery-powered alternatives. Unlike ICE vehicles, EVs are significantly more efficient. This is because they convert a higher percentage of energy into motion. They produce no exhaust, make little sound, and produce less heat than ICE vehicles. When powered with renewable electricity, they can achieve near-zero emissions when driven. Their environmental benefits depend on the energy mix of the electricity grid. In regions where coal still dominates electricity production, the carbon footprint of EVs will be higher than in areas with a clean energy grid.

The widespread adoption of EVs does present challenges. These include the need for better battery technology, expanded charging infrastructure and increased electricity generation. Lithium-ion batteries, which power most EVs, need mining for critical minerals such as lithium, cobalt and nickel. This raises concerns about resource sustainability as well as the environmental and human impacts.

Plastic and Petrochemicals

The use of petrochemicals and their contribution to greenhouse gas emissions are difficult to separate into their own category. This is because petrochemicals and other fossil fuels are intrinsic to many different sectors. For example, in industry, chemical and petrochemical use to produce energy accounts for **3.6%** of emissions [329]. Petrochemicals and other fossil fuels are also used across other sectors. For example, agriculture, cement production and aviation rely on these fuels.

Energy: Plastics fall, in part, under the energy sector as they are made from fossil fuels - crude oil and natural gas. The extraction of these is a part of the energy sector. This is because of the emissions from processes including oil drilling, natural gas extraction, and refining. Plastic production uses the same systems that make fuel for transport and electricity. So, these emissions fall under energy production and industrial processes instead of being a separate category.

Industry: Fertilisers, synthetic rubber and solvents are petrochemical products. The manufacturing of these items, along with plastics, is also embedded in the industrial sector. This is because they need energy-intensive chemical processes. This group covers emissions from cement, steel, chemicals and other manufacturing processes. However, because petrochemical production is integrated with the broader chemical industry, its emissions are not reported separately. They are all considered part of **industrial energy use and process emissions**.

Transport: Companies transport plastics and petrochemicals globally. Whether as raw materials (such as ethylene, propylene or benzene) or as finished products. Transported by ship, truck, or rail, the emissions fall within the **transportation sector** rather than the plastics category. Since producers use petrochemicals to create synthetic fuels and lubricants, they also intertwine with fossil fuel supply chains.

Waste: When discarded, plastics contribute to emissions through **landfilling, incineration, or degradation**. If incinerated, they release CO_2 **and other pollutants**. These emissions are categorised under the **waste sector** rather than plastics production. Additionally, plastics that break down in landfills can generate methane, further contributing to waste emissions.

Other Sectors: Petrochemicals are also crucial for fertilisers, pharmaceuticals, synthetic textiles, coatings and adhesives. For example, fertilisers derived from natural gas-based ammonia production contribute to agricultural emissions. Synthetic textiles, like polyester, affect the consumer goods industry and its waste management.

From raw material extraction to production, transportation, use and disposal, plastics and petrochemicals contribute to emissions at multiple points in different sectors. It is difficult to isolate them as a standalone sector. Instead, their emissions are embedded within existing sectoral categories. This makes it challenging to quantify their impact. However, as awareness of the climate impact of plastics grows, organisations are working to improve accounting methods to better track and manage petrochemical-related emissions.

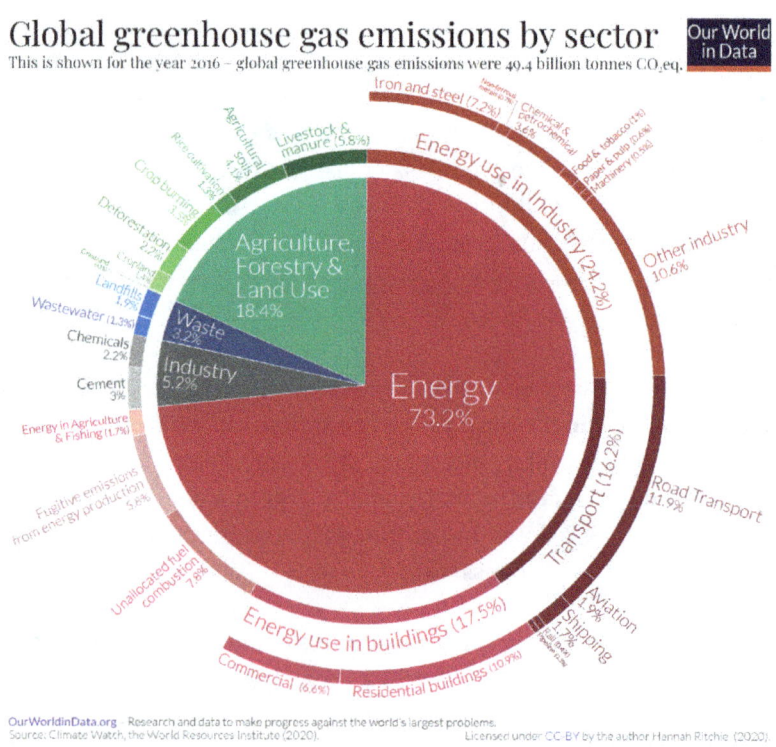

Figure 14 [327]

Agriculture, Forestry and Land Use

Agriculture

The agriculture, forestry and other land use (AFOLU) sector takes the entire food system into account. This includes **refrigeration, food processing, packaging** and **transportation.** The sector produces approximately **22% of global greenhouse gas emissions**. These emissions come from multiple sources, including livestock digestion, manure management, rice cultivation, deforestation and soil management. We have already covered many of these points in the previous chapters. The main greenhouse gases released from agriculture are methane (**CH₄**) and nitrous oxide (**N₂O**). These have a significantly higher global warming potential than carbon dioxide (**CO₂**). Methane is primarily produced by **enteric fermentation** in ruminant animals (such as cows and sheep) and by **anaerobic conditions in rice paddies**. Nitrous oxide is released from synthetic fertilisers and animal manure.

Livestock production is a major contributor to agricultural emissions, as ruminant animals produce methane during digestion. Additionally, feed production, manure storage, and processing all generate emissions. The expansion of industrial animal agriculture has increased the demand for feed crops like soy and corn. Land-use changes, clearing wild land to create crop fields, further contribute to emissions. Rice farming is another significant source. Rice grows in flooded fields. This creates ideal conditions for methane-producing microbes. Meanwhile, synthetic fertilisers applied to crops release nitrous oxide. A gas that has **300 times the warming potential of CO₂**, into the atmosphere.

Land-Use Change

Land-use change is a **major driver of carbon dioxide emissions**, through **deforestation, forest degradation and peatland destruction**. When people clear forests, grasslands, or wetlands for agriculture, the carbon stored in trees, plants, and soil is released into the atmosphere. This is a major problem in tropical regions. In these areas, vast patches of rainforest are being cut down or burnt to graze cattle or grow crops, such as soybeans and palm oil. Deforestation alone accounts for approximately **6.5% of total global emissions**, making it a critical issue for climate change mitigation.

Peatlands store vast amounts of carbon. They are especially vulnerable to land-use change. When drained for agriculture or palm oil plantations, peat releases stored carbon as CO_2. Also, soil degradation and overgrazing deplete organic matter, reducing the land's ability to absorb and store carbon. Practices like tilling release CO_2 by disturbing carbon-rich soils. Following tilling, organic matter in the soil is exposed to air. This promotes microbial activity that breaks down organic matter and releases CO_2. The more intensive the tilling, the greater the potential for CO_2 release. It has been found that tillage farming systems result in 26–31% higher carbon emissions than zero tillage systems [330].

Waste

The waste sector contributes approximately **3% of global greenhouse gas emissions**. This is primarily through the decomposition of organic waste, wastewater treatment and incineration. The main greenhouse gases released from waste management are the same as those emitted in animal agriculture. **Methane (CH_4), carbon dioxide (CO_2) and nitrous oxide (N_2O).** Methane has **80 times the warming potential of CO_2 over 20 years**, but it **remains in the atmosphere for over 100 years.** Because of this, it is the most significant concern. In the waste sector, it is produced when organic waste decomposes in landfills under **anaerobic** (low-oxygen) conditions.

Major Sources of Waste-Related Emissions

Landfills and Waste Dumps: Landfills are the largest source of waste-related methane emissions. Organic waste, such as food scraps, garden waste and paper products, produces methane when decomposing in the absence of oxygen. Unmanaged or poorly managed landfill sites release large amounts of methane directly into the atmosphere. This is common in poorer countries where waste management is not a priority. Even engineered landfills, with gas collection systems, capture only part of the methane generated. The rest escapes as emissions.

Wastewater Treatment: Wastewater from homes, industries, and agriculture contains organic matter and nitrogen compounds. When it decomposes, it releases methane and nitrous oxide. In **anaerobic conditions** (such as in untreated or poorly treated sewage), these releases increase. Wastewater treatment plants can also produce CO_2 from the energy-intensive treatment processes.

Incineration and Open Burning: Burning any material releases CO_2. But when waste is burnt, **black carbon** can result. This is a short-lived climate pollutant that contributes to global warming. Burning plastic waste can release toxic pollutants like dioxins and furans. This creates concerns for both public health and the climate. While modern waste-to-energy plants capture some emissions to generate electricity, they still contribute to CO_2 emissions. This is made worse when burning fossil fuel-derived materials like plastics to heat water to generate steam for turbines.

The Ghazipur Landfill, India

The **Ghazipur landfill** is one of India's largest and most problematic landfills. It is located in the village of Ghazipur, in the eastern district of Delhi, India. They established the landfill site in 1984 and, despite reaching maximum capacity in 2002, it continues to receive waste from Delhi [331].

The landfill covers an area of approximately 70 acres (28 ha) and has a height of over 72 m (236 feet). It is **one of the largest landfills in the Delhi region**, where waste management is poorly regulated. One of the hazards created through the lack of management is fires. Fires are frequent on the "garbage mountain" of more than 14 million metric tonnes of waste. The burning waste spreads smoke and toxic fumes across the residential and commercial areas nearby.

> *"Its stench is nauseating, breathing is difficult and a toxic taste engulfs the mouth and throat." - Neville Lazarus, Sky News*
> [332]

There have also been several collapses. In September 2017, two people died when 50 million tonnes of waste broke away. The mound of waste landed in water, creating a wave that washed cars and motorbikes from a road nearby into a drain. Rescuers retrieved five people from the water, but two of them died before rescue. Those who saw the incident reported hearing a loud explosion, from the sudden release of built-up gases. There was then a huge wave of water and waste that hit vehicles on the road [331].

Landfills, in general, release a variety of **greenhouse gases and toxic pollutants**. They contribute to air pollution, climate change and create health hazards. However, the situation in Ghazipur is particularly bad. One doctor working at a local hospital, Dr Arshad

Khan, spoke to Sky News about his experience of trying to help his patients. "It's a slow poisoning of people living here. The poison enters their bodies through the air, weakening their immunity, and medicines are not as effective as they should be" [332].

Kayrros, an energy and environmental intelligence company, has collected satellite data over the landfill. It shows methane emissions with an average flow rate of 2.91 tonnes per hour between November 2021 and May 2024 [333]. Some consider this rate to be on the low side. But, if this flow rate is constant, the amount of methane released each year would amount to approximately 25,000 tonnes. This is equivalent to the 20-year warming impact of CO_2 emissions from 500,000 cars running simultaneously the whole year round.

Water pollution from the Ghazipur site is another very serious environmental problem. Dark streams of fluid drain from the landfill into the earth, contaminating the soil, groundwater and surface water. A canal that runs parallel to the waste pile has become putrid and completely blackened with contamination. Plastic waste polluted the water. Animal body parts, including fish and chicken innards from slaughterhouses, float in the water.

What Can We Do?

Large-scale systemic changes are necessary to tackle climate change. But individual actions **do add up** and can drive demand for broader changes. There are direct actions that we can all take. We can reduce water use and food waste. We can dispose of unavoidable food waste, such as onion peels and carrot stalks, by composting or using food waste collection. The key is to prevent food waste from being sent to landfill. We can reduce energy use with efficient, low-energy appliances. Unplug any device that we are not using, including chargers, and choose energy providers offering renewable electricity. More expensive options include installing solar panels. Houses can get cavity insulation and energy-efficient windows, reducing the need for active heating and cooling. There are also heat pumps that use the Earth's energy to heat buildings, instead of fossil fuels.

We can reduce our transport emissions by walking, riding a bike and using public transport whenever possible. If driving is unavoidable, consider carpooling or switching to an electric vehicle. Also, minimise plane trips as much as possible. Additionally, we can be more mindful consumers by reducing plastic use. Recycle and buy second-hand, whether in vintage shops, charity shops or on websites such as eBay or Vinted. In addition to these choices, we can use ethical banking services that invest in sustainable and renewable energy sources. Buy from companies with strong environmental policies and avoid those with high carbon footprints. The many small changes that we can make are powerful when we work together.

We can take indirect actions. Talking about climate change and supporting politicians with policies that prioritise climate action can put these issues on the political agenda. However, many of the solutions that would have a real impact on the emissions from energy use, waste and the chemical industry are beyond our control. These matters are driven by the market. For example, a company may decide to refit their office space with LED lights with timed sensors. These devices only turn on when people enter the space and turn off if no movement is detected. New buildings can be constructed with high-quality insulation, solar panels and a green roof of grass or other plants. A delivery company may replace their fleet with electric vehicles when the lease renewal is due. We, as individuals, have no impact on these decisions. But a business may feel that there is marketing potential in appearing to be more "green". They may reduce plastic in packaging or switch to paper; they may even make claims of planting trees or offsetting their carbon emissions in other ways. These strategies are called "greenwashing." It's a marketing tactic that makes a company or product seem eco-friendly. However, it doesn't actually reduce environmental impacts.

Other actions that would make a significant difference to the trajectory of climate change would be to end deforestation. Reforest areas that have been cleared. Plant new forests on open ground. Rewilding projects, such as the ones in the Netherlands that we looked at in chapter 4, are already having a positive impact on the local environment. This is definitely the case when compared to the negative impact of the activities that were taking place on those same areas of land.

Significantly reducing industrial emissions is not something that individuals can achieve. These changes are on a scale that requires whole sectors to revise their activities. They need to install different methods of operation. Employ new technologies, such as carbon capture and storage, green hydrogen production and circular use economy initiatives. All of these actions are vital.

Governments must also act, assessing opportunities to implement climate-positive measures. Policy changes that allow for more renewable energy development. This could include giving out grants and government investments, or allocating land for wind turbines and solar panels. Governments can also encourage corporations to embrace less polluting policies with the use of carbon taxes or financial incentives. Local governments, making decisions on planning applications, can require certain specifications. Buildings may need efficient insulation, and they can state that trees must be planted as part of any new development. But again, these are not things that you or I can do. Unless we are the CEO of a very successful multinational company or the leader of a country, we cannot make any changes to these systems that will result in significant change for the planet.

Conclusion

Almost **75% of global greenhouse gas emissions come from energy use**, including electricity generation, transportation and industry. These emissions are deeply embedded in the infrastructure that powers our world. We rely on fossil fuel-based energy grids, manufacturing systems and supply chains designed around coal, oil and gas. Transitioning to renewable energy, electrifying transportation and improving efficiency are all essential for reducing emissions. But these systemic changes need government policies, corporate investment, technological advancements and, most of all, money. As individuals, our ability to influence these sectors is limited. We cannot redesign power plants, overhaul factories, or dictate energy policies. However, choosing to buy sustainable items and services can create demand-side stimulus. Personal actions, like reducing energy use or driving less, can make a difference. But they are small compared to the structural changes needed to decarbonise these sectors. This is why it is so important to address the changes that we can make, such as what we eat.

Unlike energy grids or industrial supply chains, **what we eat is a choice we make every day**. A choice that directly impacts emissions. Animal agriculture is one of the largest contributors to climate change. Yet it exists and operates at its current scale **because of demand**. Shifting away from animal-based foods is one of the most powerful ways an individual can reduce their environmental impact. Reducing their carbon footprint, conserving water, and freeing up land for reforestation. Unlike other sectors, where emissions are locked into vast infrastructures, food is a system that **we have the power to change**. We simply need to make a conscious choice about what we put on our plates.

Chapter 11: Putting it All Together

The previous chapter ended on a rather negative note. Most of the greenhouse gas releases driving climate change result from processes that we cannot control. This does not mean that we can do nothing. By making conscious choices about what we eat and drink, we can greatly reduce the emissions that are warming our planet. To the tune of 25% by some estimates. A report from WWF-UK states that even "Halving UK meat consumption, to approximately WHO recommended levels, could save roughly 10 MtCO$_2$e" [334]. This means eating less meat, following the World Health Organization's healthy guidelines, instead of overindulging like many do now.

Veganism

Going vegan significantly reduces resource consumption and our environmental impact. For those who do not want to go vegan, reducing the amount of animal products in their diets would still benefit the planet. The impact increases as animal products are reduced. It is possible to approximate the benefits, or savings, of a vegan diet. The figures are approximate due to variations in climate, geography, infrastructure and regulations in different countries.

Below is a breakdown of key resources saved per day, per month and per year by an individual going vegan:

1. Greenhouse Gases (CO_2 & Methane) Saved
    - **Per day**: ≈10 kg CO_2 equivalent
    - **Per month**: ≈300 kg CO_2 equivalent
    - **Per year**: ≈3,600 kg CO_2 equivalent

Equivalent to driving about **5,900 miles (9,500 km)** in an average internal combustion engine car.

2. Methane Reduction (CH_4)
    - Animal agriculture causes 37% of methane emissions, a gas with **80x the warming potential of CO_2** over 20 years. Put another way, **going vegan means a 37% reduction in methane emissions**.

3. Water Saved
    - **Per day**: ≈ 4,200 litres (1,100 gallons)
    - **Per month**: ≈125,000 litres (33,000 gallons)
    - **Per year**: ≈1.5 million litres (400,000 gallons)

Equivalent to not showering for **66 years**, an unpleasant comparison, but true.

4. Land Saved
- **Per day**: ≈2.8 m² (30 square feet)
- **Per month**: ≈84 m² (900 square feet)
- **Per year**: ≈1,000 m² / 0.1 hectare (10,950 square feet)

Equivalent to the size of **two tennis courts**.

5. Animal Lives Saved
- **Per day**: ≈1 animal
- **Per month**: ≈30 animals
- **Per year**: ≈365+ animals

This includes land animals and an even greater number of marine animals (due to bycatch in fishing). People have stated that if you eat a piece of an animal, a chicken leg, or a salmon steak, you are not eating a whole animal. But to eat that piece of chicken or salmon, the whole animal has to die.

6. Forest & Biodiversity Preservation
- 91% of **Amazon deforestation** is driven by animal agriculture.
- Going vegan reduces the demand for land used for livestock and feed crops, which helps protect forests, wildlife and indigenous lands.

7. Ocean & Waterway Protection
- Animal agriculture is **the leading cause** of ocean dead zones, water pollution and species extinction, due to runoff of manure, antibiotics and fertilisers.
- **80% of global fish stocks** are overfished or depleted. Switching to a plant-based diet reduces the demand for destructive fishing practices.

8. Antibiotic Reduction
- **70–80% of antibiotics worldwide** are used in animal farming, leading to antibiotic-resistant bacteria. This applies mainly to the farming of land animals, but also aquaculture.

- Going vegan helps reduce antibiotic use and the risk of antibiotic resistance in humans.

These numbers are not insignificant. Per person, going vegan has a huge cumulative impact over time, saving **thousands of animals, millions of gallons of water and tonnes of CO_2**, while reducing deforestation, pollution and disease risks.

Individual and Systemic Actions

Shifting toward plant-based diets is a key strategy for reducing agricultural emissions. But there are many other approaches to cutting greenhouse gas (GHG) production across different sectors. These methods require the actions of individuals, industry and governments working together with the common goal of reducing emissions and pollutants.

Potential GHG Emission Reductions from Individual and Systemic Actions

Estimating the impact of individual actions on global GHG emissions is complex. Any reductions depend on **widespread adoption, regional differences, and technological advancements**.

It is possible to make rough estimates based on current research and sectoral emission breakdowns:

1. Transitioning to Renewable Energy (≈25-30% reduction)

- Energy production accounts for **73% of global GHG emissions**, with electricity and heat generation making up 30-35%.

- Switching to **100% renewable energy** for electricity could cut emissions by **25-30%** globally. It is also important to note that there is an amount of energy required to produce energy. An example is boiling water for steam to turn turbines that generate electricity. Renewable energy sources, such as wind turbines, reduce energy use by not needing the initial energy inputs. Another example is the energy used to drill for and transport gas and oil used for fuel. The use of solar power and heat pumps would make these processes redundant.
- Individual actions like using renewable energy providers and solar panels, and reducing electricity use, have a **small direct impact**. But they also influence systemic change through demand-side stimulus, shaping markets.

2. Electrifying Transportation (≈10-15% reduction)

- Transportation accounts for **16-20%** of global emissions.
- Shifting to electric vehicles (EVs) and expanding public transport could **cut 10-15%** of total emissions.
- Reducing car travel, taking public transport, biking, and flying less can contribute to this reduction.

3. Improving Energy Efficiency in Buildings (≈5 - 10% reduction)

- Buildings (residential & commercial) contribute **17-18%** of emissions.
- Retrofitting buildings, improving insulation, using heat pumps and switching to energy-efficient appliances could cut **5-10%** of total emissions.
- Individuals can reduce energy use at home with efficient use of appliances. They can do larger loads in the washing machine, so they use the machine less, saving energy and water. Installing efficient appliances is also effective.

4. Reducing Waste (≈5% reduction)

- Waste management contributes **3-5%** of global emissions, mostly from landfill methane.
- Reducing food waste and improving recycling and composting could nearly **eliminate landfill emissions**. This would cut **up to 5%** of total GHGs.

5. Carbon Sequestration Through Reforestation & Land Protection (≈10-15% reduction)

- Deforestation and land use changes contribute **15-18%** of emissions.
- Ending deforestation, reforesting land, and protecting natural carbon sinks could remove **10-15%** of emissions.
- Supporting reforestation efforts and reducing consumption of products linked to deforestation (e.g., palm oil, soy and timber) can help.

6. Sustainable Consumer Choices & Circular Economy (≈5% reduction)

- Reducing material consumption, buying second-hand and recycling materials more effectively could cut emissions in manufacturing and waste disposal.
- Extending product life cycles, reducing fast fashion and using durable goods can **reduce emissions by ≈5%** globally.

Estimated reduction potential:

- Renewable energy: 25-30%
- Transportation electrification: 10-15%
- Building efficiency: 5-10%
- Waste reduction: 5%
- Reforestation & land protection: 10-15%
- Consumer behaviour & circular economy: 5%

What's Realistically Achievable?

While individuals can contribute, **systemic policies, corporate action, and government regulations are critical**. It is possible to **cut GHG emissions by 40-50% by 2050**. For this to happen, governments must enforce strong climate policies while industries adopt clean technologies. For maximum effect, individuals should **focus on high-impact changes such as energy, transport and advocacy**. We also need to push for systemic change through **voting, activism and influencing businesses**.

Sources:

- IPCC Special Report on Global Warming of 1.5°C (2018) [3]
- IPCC Sixth Assessment Report (AR6), Working Group III: Mitigation of Climate Change (2022) [335]
- Project Drawdown: The Drawdown Review (2020) [336]
- IEA Net Zero by 2050: A Roadmap for the Global Energy Sector (2021) [337]
- McKinsey & Company: The Net-Zero Transition (2022) [338]
- FAO: Food Wastage Footprint – Impacts on Natural Resources (2013) [339]
- Griscom et al.: Natural Climate Solutions (2017) [340]
- WWF & Imperial College London: Behaviour Change, Public Engagement and Net Zero (2020) [341]

Potential GHG Emission Reductions from Eliminating Animal Agriculture

Animal agriculture causes **14.5-20%** of global greenhouse gas (GHG) emissions.

This includes emissions from:

- methane (CH_4)
- deforestation
- land use change
- feed production
- manure management
- energy use in livestock farming.

Eliminating animal agriculture could lead to **a total reduction of up to 20% in global GHG emissions**. But the precise number depends on **how land is repurposed** after removing livestock.

Breakdown of Animal Agriculture Emissions		
Source	% of Global Emissions	Reduction Potential
Methane from enteric fermentation (cow burps)	4-5%	Near-total elimination
Manure management & emissions	2%	Near-total elimination
Feed production (fertilisers, land use, processing, transport)	3-5%	Major reduction
Deforestation for pasture/feed crops	4-6%	Major reduction
Energy use in animal agriculture (transport, slaughter, processing, refrigeration, etc.)	2%	High reduction
Total Impact of Eliminating Animal Agriculture	15-20%	Realistic reduction: ≈12-15% (considering transition costs)

Table 11

When comparing the possible total emission reduction from a vegan lifestyle to other emission reduction actions, a vegan lifestyle is equivalent to electrifying transport plus improving the energy efficiency of buildings.

Potential Reductions Through Other Climate Actions	
Action	Potential GHG Reduction
Eliminating animal agriculture	15-20%
Transitioning to 100% renewable energy	25-30%
Electrifying transport (EVs, public transit, etc.)	10-15%
Improving building energy efficiency	5-10%
Reforestation & land protection	10-15%
Reducing waste & transitioning to a circular economy	5-10%

Table 12

The impact that a vegan lifestyle can have on greenhouse gas emissions is impressive when you consider that this is something that we can do now. We don't need the government to step in. We don't need planning permission, investment, infrastructure, or equipment. A recipe book or two could help, but beyond this, we are fully equipped to do this today.

Mental and Emotional Suffering

When talking about animal agriculture, the focus is almost entirely on the animals and rightly so. Every year, humans use billions of animals in ways that are unethical and cruel, only to satisfy desires. This is wrong. But it is also important to discuss the suffering of those who work with those animals. The breeders, farm workers, slaughterhouse workers, and the staff of processing plants. For example, those who work in the egg industry, killing hundreds of newly hatched chicks a day, suffer from the required horror of their jobs. But this is not talked about. Those who eat the products that come from these processes of breeding, raising, and killing don't consider the human harm. Those who object to the processes of the use of animals rarely care for the humans involved. Some even feel that these people deserve to suffer for what they have done.

Mental Health Issues Among Farmers and Agricultural Workers

Depression and Anxiety: Mental health issues, like depression and anxiety, are common among farmers. This is due to stress from financial pressures, long working hours, and social isolation. In some cases, physical illness related to the demands of the job adds to their strain. Studies on mental health in rural populations often highlight that farmers experience significantly higher rates of depression and anxiety compared to the general population. A study, published in *Frontiers in Public Health* in 2020, investigated this. It found that agricultural workers had higher rates of depression than the general U.S. population. Rates were estimated to be up to **50%** among farmers, compared to about **17%** of the general population [342]. Anxiety disorders are also common, particularly among farmers experiencing financial instability or dealing with unpredictable weather.

Post-Traumatic Stress Disorder (PTSD): Slaughterhouse workers, out of all groups working in animal agriculture, are at a notably elevated risk of **PTSD**. The highly stressful and often traumatic nature of their work causes them serious mental health problems.

This includes exposure to violence, handling and slaughtering animals. They will often witness accidents or injuries to their colleagues. The emotional toll can be profound, especially for those directly involved in animal slaughter. According to a study published in the *Journal of Occupational Health Psychology* (2013), **15%-20%** of slaughterhouse workers exhibited symptoms of PTSD. Their symptoms include flashbacks and emotional numbness [343].

Substance Abuse: Both agricultural workers and slaughterhouse workers are more likely to use substance abuse as a coping mechanism. Rates of alcoholism are high among slaughterhouse workers. As is the use of illegal substances. These behaviours result from the need to manage the stress that they suffer. They deal with many mental health issues from the high-pressure environment of mass killing animals. Workers often feel isolated. They have limited access to mental health resources. They also feel unable to discuss their problems with those outside of the industry. This includes loved ones, as this is not a job that the public can relate to.

> *Prevalence of serious psychological distress (SPD) among workers was 4.4%, compared to United States population-wide prevalence of 3.6%. Prevalence of mild and moderate psychological distress among these workers (14.6%) was also higher than national estimates. - Jessica H Leibler et al* [344]

Suicide

In the United States, the suicide rate among farmers is notably higher than that of the general population. Data from the Centers for Disease Control and Prevention (CDC) states that farmers have suicide rates roughly 1.6 times higher than the national average. Suicide rates among farmers and agricultural workers are a significant issue in many countries.

USA: A 2023 report from the CDC said farmers, ranchers and other agricultural workers had a suicide rate of about 49.9 per 100,000 individuals. This is compared to the male suicide rate of 32.0 per 100,000 [345].

Australia: In Australia, a 2012 study found the suicide rate among farmers was approximately 1.7 times higher than the general population [346]. The Australian Broadcasting Corporation (ABC) reported in 2020 that this was due to factors including financial stress, droughts, and mental health challenges.

Canada: The review article, *Understanding the factors contributing to farmer suicide: a meta-synthesis of qualitative research,* 2023, looked at farmer suicides. It found that farmers are at a higher risk of suicide compared to the general population. Risk factors included financial strain, isolation and the high-stress nature of farming. This was particularly true in rural and remote areas [347].

Europe: Across Europe, including the UK, France and Germany, many countries have reported higher rates of poor mental health and suicides among those working in animal agriculture, compared to the general population. Common factors across nations include financial hardship, isolation and high workloads. Financial strain is common, as although people want to consume the products of animal agriculture, they do not want to pay a great deal for them. Similarly, few people want to work in animal agriculture, and no one wants to live near a slaughterhouse. These factors mean that workers, including farmers, make little money from a gruelling job. A job that requires them to work away from the rest of society, and feel unable to talk about the work stress they suffer from [348].

Themes Within Animal Agriculture

When we examine the environmental damage caused by animal agriculture, we see a pattern. The same pollutants and emissions appear over and over again. Greenhouse gases like **methane**, **carbon dioxide** and **nitrous oxide**. Water pollutants such as **nitrogen**, **phosphorus** and **antibiotics**. The industry's footprint is remarkably consistent and overwhelmingly destructive. These emissions seep into the air we breathe, the water we drink and the land we depend on. They create a cascading series of environmental crises that are becoming impossible to contain.

Methane, primarily released by ruminant animals like cows and sheep, is one of the most insidious emissions. Though it doesn't linger in the atmosphere as long as carbon dioxide, its warming potential is over 80 times greater in the short term. It is accelerating climate breakdown at an alarming rate. Meanwhile, the vast amounts of manure produced by factory farms release a multitude of pollutants. They include nitrous oxide, another potent greenhouse gas. At the same time, manure contaminates water sources with excessive nutrients. These pollutants don't disappear. They accumulate in the environment. They fuel algal blooms, creating dead zones and rendering freshwater supplies undrinkable.

What makes these emissions particularly dangerous is their persistence and interconnectedness. Air pollution from factory farms exacerbates respiratory illnesses, especially in vulnerable communities living nearby. Runoff from animal waste and synthetic fertilisers poisons aquatic ecosystems. Biodiversity is lost as animals die or move away. Grazing animals and the growing of feed crops are causing deforestation. The process releases stored carbon into the atmosphere, while stripping the planet of its ability to absorb future emissions. In essence, the same pollutants - methane, carbon dioxide, nitrous oxide and nutrient runoffs - keep surfacing in different but devastating ways.

By recognising these patterns, it becomes clear that animal agriculture is not just harming the environment through isolated processes. It is systematically degrading every major life-supporting system on Earth. The surreptitious nature of these emissions makes them easy to overlook in day-to-day life. But their consequences are undeniable. With every passing year, they erode the planet's ability to recover, destabilising ecosystems and intensifying climate extremes. If left unchecked, if we continue as we are, they will push the Earth past critical tipping points. The damage will be irreversible. It will be damage that no amount of intervention can undo.

Reducing Water Use and Pollution

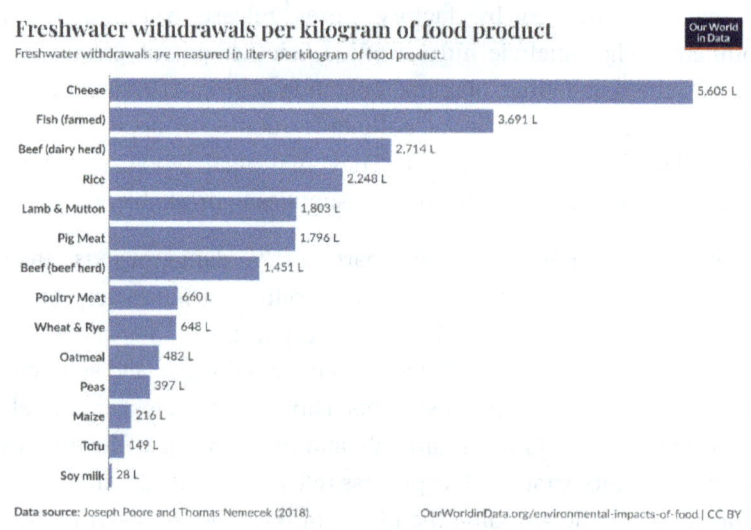

Figure 1

One of the most significant yet often overlooked environmental benefits of going vegan is the drastic reduction in freshwater use. The livestock industry is one of the largest consumers of freshwater on the planet. It uses vast amounts of water, not only for the animals but also for farm processes and growing their feed. For example, producing one pound of beef needs anywhere from 1,800 to 2,500 gallons of water. Growing crops like lentils, beans, and potatoes need only a fraction of that amount. By shifting away from animal agriculture and towards plant-based foods, we can dramatically decrease global freshwater withdrawal. This would ensure that this essential resource remains available for human consumption and ecosystems.

In addition to water conservation, vegan diets also significantly reduce water pollution. Factory farms and feedlots generate immense amounts of animal waste. This often ends up in groundwater and waterways. In the water, they cause harmful algal blooms. Oxygen-depleted water leads to dead zones. Nutrient overload, pathogens and chemicals cause contamination of drinking water supplies. The excessive use of fertilisers and pesticides for growing animal feed crops further exacerbates this problem. Nitrate pollution poisons water sources and harms both humans and wildlife. In contrast, growing plant-based foods minimises these risks, keeping our rivers, lakes, and oceans cleaner.

By choosing a vegan lifestyle, individuals can play a direct role in protecting our planet's freshwater systems. Fewer animals raised for food means less water waste, less pollution and healthier ecosystems. Water scarcity is an increasing concern worldwide. Shifting to plant-based eating is one of the most effective ways to preserve this resource for today and future generations.

Ethics

Beyond its environmental consequences, animal agriculture raises serious ethical concerns. Humans breed billions of animals into existence, only to confine, exploit and slaughter them. They are not treated as sentient beings, but as units of production. In factory farms, mother cows are forcibly separated from their calves. Pigs live in crates so small they cannot turn around. Chickens, selectively bred, grow so unnaturally fast that their legs often collapse under their weight. These are not isolated incidents. They are routine industry practices, normalised only because they happen behind closed doors.

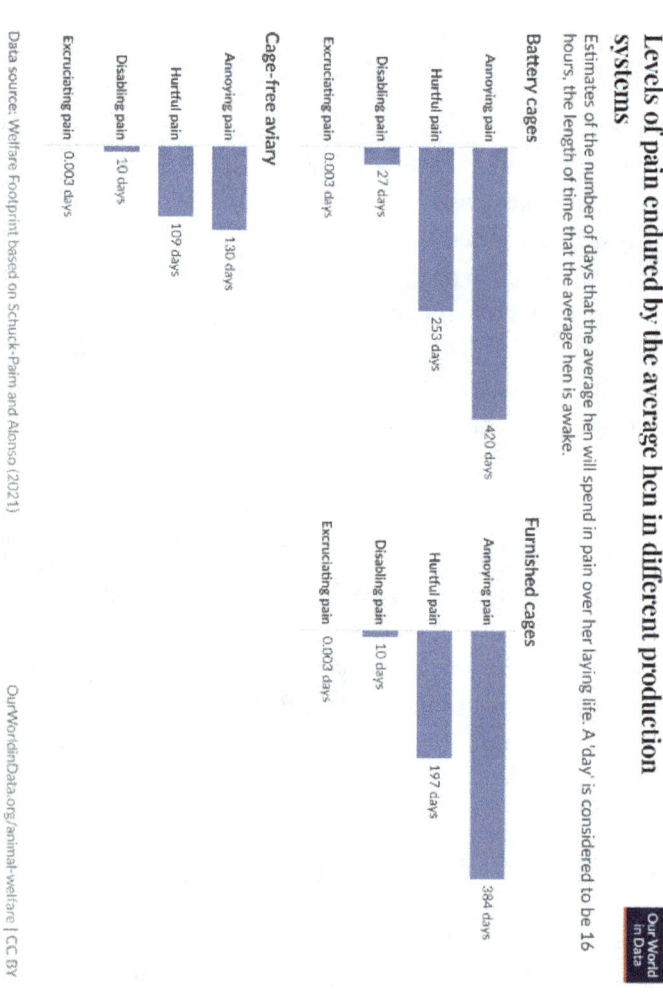

Figure 15

Figure 15 above from Our World in Data, "Levels of Pain Endured by the Average Hen in Different Systems", reveals a stark disparity in animal welfare across production methods. Hens confined in battery cages endure extremely high levels of pain. They suffer from extreme limitations on movement. They live in persistent overcrowding. They endure the cumulative stress of an environment designed purely for productivity. Even in systems marketed as more humane, such as free-range or organic, the pressures of industrial farming still cause considerable distress to these animals. This visual evidence forces us to confront the troubling reality. Regardless of the production system, the ethical cost, relating to animal suffering, is both measurable and profound.

If someone inflicted these same conditions on a pet, child, or family member, the outcry would be immediate and overwhelming. No one would accept a newborn being taken from its mother. A child confined in a cage too small to move in. No one would pay a person who causes their family pet suffering simply for convenience. Yet, when it comes to animals in the food system, these realities are ignored or excused. The ethical question is not whether we *can* continue this system, but whether we *should*.

Conclusion

We stand at a cliff edge – the precipice of destruction – where the choices we make now will decide whether life on Earth recovers or collapses. Animal agriculture is one of the greatest threats to the survival of our planet. It is the main cause of deforestation. It destroys forests that regulate our climate and provide shelter for many species. It devours freshwater at unsustainable rates, draining the rivers that millions rely on. It causes almost a quarter of greenhouse gas emissions, heating the atmosphere and disrupting the climate. Animal agriculture poisons our soil, rivers, and oceans with waste, pharmaceuticals and chemicals. Natural bodies of water become fetid, lifeless dead zones. These are not isolated problems but interconnected crises. Symptoms of a food system fundamentally at odds with the planet's ability to sustain life.

But it is not only the Earth itself that suffers. Farming traps billions of animals in a cycle of exploitation. The industry breeds them into existence to endure confinement, mutilation, and early death. Humans, too, caught in this system, are suffering. Workers face dangerous, toxic conditions and low wages. In slaughterhouses, people endure the mental anguish of killing day after day. Communities living near factory farms breathe polluted air and drink contaminated water. They experience higher rates of illness, becoming victims of asthma, skin disorders and cancer. At the same time, livestock consume vast quantities of food while millions of people go hungry. Animal agriculture is not simply destructive – it is profoundly unjust. It indiscriminately inflicts harm on animals, ecosystems, and human beings alike.

The health costs are no less stark. Diets heavy in meat, dairy, and eggs increase the risk of heart disease, cancer, and diabetes. Plant-based diets lower those risks while promoting vitality. People see veganism as a sacrifice, but this is not the case. Veganism is a chance to nourish ourselves better, to extend life spans, and to prevent unnecessary suffering. It is also a path toward justice. The grains and legumes now used to fatten animals could feed billions of people, creating plenty rather than scarcity.

We cannot ignore these truths. To continue down the current path is to walk willingly into collapse. But there is another way. By rejecting animal agriculture, we can protect the planet, safeguard human health, and honour the dignity of all life. This is the legacy we owe future generations: not a barren, broken world, but one where people, animals, and ecosystems thrive together. The time to step back from the brink is now.

Figures

1. Water use per kg of food product
2. Land use per 100g of protein
3. Global land use for food production
4. Land use by diet
5. Eutrophication by kg of food
6. Agricultural freshwater use as a share of total water withdrawal
7. Meat consumption per person
8. Land use in aquaculture
9. Freshwater use in aquaculture
10. Greenhouse gas emissions from aquaculture
11. Deforestation by cause
12. Environmental impacts of dairy
13. Global Greenhouse Gas Emissions by Sector
14. Global greenhouse gas emissions by sector
15. Levels of Pain Endured by the Average Hen in Different Systems

Tables

1. Water consumption in pig farming
2. Water consumption in cattle farming
3. Water footprint for $2m^2$ of fabric
4. Water consumption: leather vs other fabrics
5. Energy requirements MJ per kg of fabric
6. The pollution and inefficiency of animal leathers
7. Animal leathers and plant-based alternatives: side by side
8. Cost comparison: animal leather vs plant-based alternatives
9. Environmental impacts of fur
10. Properties comparison: animal leather vs plant-based alternatives
11. Breakdown of animal agriculture emissions
12. Potential reductions through other climate actions

References

1. **Mulhern, O.** A Graphical History of Atmosphere CO2 Levels Over Time. *Earth.org.* [Online] August 2020. https://earth.org/data_visualization/a-brief-history-of-co2/.

2. **NASA.** What is Climate Change. *NASA.gov.* [Online] https://science.nasa.gov/climate-change/what-is-climate-change/.

3. **(IPCC), Intergovernmental Panel on Climate Change.** Global Warming of 1.5°C: An IPCC Special Report on the impacts of global warming of 1.5°C above pre-industrial levels and related global greenhouse gas emission pathways. *IPCC.* [Online] 2018. [Cited: 21 March 2025.] https://www.ipcc.ch/sr15/.

4. Remarques Generales sue les Temperatures Du Globe Terrestre et des Espaces Planetaires. **Fourier, J.** s.l. : Annales de chimie et de Physique (in French), 1824.

5. Circumstances affecting the Heat of the Sun's Rays. **Foote, Eunice.** 1856, Vol. 22.

6. Happy 200th birthday to Eunice Foote, hidden climate science pioneer. **Huddleston, Amara.** s.l. : NOAA, 2019.

7. Carbon Dioxide Exchange between Atmosphere and Ocean and the Question of an Increase of Atmospheric CO2 During the Past Decades. **Revelle, Roger, and Hans E. Suess.** s.l. : Tellus, 1957.

8. **Counts, The World.** Global Warming. *The World Counts.* [Online] 2024. https://www.theworldcounts.com/challenges/climate-change/global-warming.

9. H.-O. Pörtner, D.C. Roberts, V. Masson-Delmotte, P. Zhai, M. Tignor, E. Poloczanska, K. Mintenbeck, A. Alegría, M. Nicolai, A. Okem, J. Petzold, B. Rama, N.M. Weyer (eds.). *IPCC Special Report on the Ocean and Cryosphere in a Changing Climate.* s.l. : IPCC, 2019.

10. **NASA.** Ocean Warming | Vital Signs - Climate Change. *NASA.gov.* [Online] https://climate.nasa.gov/vital-signs/ocean-warming/.

11. **Office, Met.** UKCP18 Guidance: Representative Concentration Pathways. s.l. : Met Office, 2018.

12. **Brief, Carbon.** Analysis: Just four years left of the 1.5C carbon budget. s.l. : Carbon Brief, 2017.

13. **Fetch, Sarah.** State of the Planet. *Columbia Climate School.* [Online] 2020. https://news.climate.columbia.edu/2020/01/27/wine-regions-shrink-climate-change/.

14. **Hansen, James.** Greenhouse Effect and Global Climate Change. Washington D.C. : s.n., 1988.

15. **Simmons, Daisy.** Myth: Why two degrees of global warming is worse than it sounds. *Yale Climate Connections.* [Online] 8 February 2023. https://yaleclimateconnections.org/2023/02/myth-why-two-degrees-of-global-warming-is-worse-than-it-sounds/.

16. Climate change: What happens if the world warms up by 2°C? Sky News; YouTube, 2019.

17. Summit, UN Climate. Comparing climate impacts at 1.5°C, 2°C, 3°C and 4°C. *UN Climate Summit.* [Online] https://unclimatesummit.org/comparing-climate-impacts-at-1-5c-2c-3c-and-4c/.

18. NASA. NASA study reveals compounding climate risks at two degrees of warming. *NASA.* [Online] 2021. https://www.climate.nasa.gov/news/3278/nasa-study-reveals-compounding-climate-risks-at-two-degrees-of-warming/.

19. Heat tolerance in plants: An overview. Environmental and Experimental Botany, 61(3), 199 223. Wahid, A., Gelani, S., Ashraf, M., & Foolad, M. R. s.l. : Elsevier, 2007.

20. Safdie, Stephanie. What Would be the Impact of a 3°C Rise in Global Temperature? *Greenly.* [Online] 28 November 2023. https://greenly.earth/en-us/blog/ecology-news/what-would-be-the-impact-of-a-3degc-rise-in-global-temperature.

21. Economist, The, [prod.]. See what three degrees of global warming looks like. YouTube, 2021.

22. What will our world look like at 4 degrees? PBS Terra; YouTube, 2024.

23. *The interaction of climate change and methane hydrates.* Carolyn D. Ruppel, John D. Kessler. 1, s.l. : AGU, 2016, Vol. 55.

24. University, Wageningen. Past extreme climate warming triggered by tipping points, study finds. *Phys.org.* [Online] 2023. https://phys.org/news/2023-04-extreme-climate-triggered.html.

25. *The impacts of hydrate mining.* Review, World Ocean. s.l. : World Ocean Review, 2014.

26. Planet 6C: Will climate change turn planet Earth into Mars? *Climate Change News.* [Online] Climaye Change News, 2012. https://www.climatechangenews.com/2012/11/08/planet-6c-will-climate-change-turn-planet-earth-into-mars/.

27. Svitek, Patrick. Texas puts final estimate of winter storm death toll at 246. *The Texas Tribune.* [Online] 2023. https://www.texastribune.org/2022/01/02/texas-winter-storm-final-death-toll-246/.

28. Paws, Four. The Life Expectancy of Farm Animals. *Four Paws.* [Online] 2 April 2024. https://www.four-paws.org/campaigns-topics/topics/farm-animals/age-of-farm-animals.

29. Jackson, Joe. Gone in 30 years? The Welsh village in crosshairs of climate change. s.l. : Phys.org, 2022.

30. Gerretsen, Isabelle. The Welsh village of Fairbourne is at high risk of being submerged from rising sea levels and residents have been told they will need to move. But many say they will refuse to leave. *BBC Future Planet.* [Online] 10 May 2022. https://www.bbc.co.uk/future/article/20220506-the-uk-climate-refugees-who-wont-leave.

31. Press, The Associated. Portugal declares a state of calamity as wildfires rage out of control. *NPR.* [Online] 19 September 2024. https://www.npr.org/2024/09/19/g-s1-23693/portugal-wildfires-rage-out-of-control.

32. Adam Easton, Malu Cursino, Ruth Comerford. 'Catastrophe' as deadly floods hit Central and Eastern Europe. *BBC News.* [Online] 15 September 2024. https://www.bbc.co.uk/news/articles/c0jwp3ppp6xo.

33. DW. Slovakia's capital hit by biggest floods in 30 years. *Times of Oman.* [Online] 19 September 2024. https://timesofoman.com/article/150030-slovakias-capital-hit-by-biggest-floods-in-30-years.

34. AP. European Union warns deadly flooding, wildfires show climate breakdown fast becoming the norm. *The Telegraph Online.* [Online] 18 September 2024. https://www.telegraphindia.com/world/european-union-warns-deadly-flooding-wildfires-show-climate-breakdown-fast-becoming-the-norm/cid/2048773.

35. Attribution, World Weather. Deadly Mediterranean heatwave would not have occurred without human induced climate change. *World Weather Attribution.* [Online] 31 July 2024. https://www.worldweatherattribution.org/deadly-mediterranean-heatwave-would-not-have-occurred-without-human-induced-climate-change/.

36. Siegfried, Kristy. Climate change and displacement: the myths and the facts. *UNHCR The UN Refugee Agency.* [Online] 15 September 2023. https://www.unhcr.org/news/stories/climate-change-and-displacement-myths-and-facts.

37. Turrentine, Jeff. Climate Misinformation on Social Media Is Undermining Climate Action. *NRDC.* [Online] 19 April 2022. https://www.nrdc.org/stories/climate-misinformation-social-media-undermining-climate-action.

38. Shahan, Zachary. Why Arnold Schwarzenegger Wants Us To Focus On "Pollution" Rather Than "Climate Change". *Clean Technica.* [Online] 2023. https://cleantechnica.com/2023/06/06/why-arnold-schwarzenegger-wants-us-to-focus-on-pollution-rather-than-climate-change/.

39. McGrath, Matt. Climate change: 'Clear and unequivocal' emergency, say scientists. *BBC News.* [Online] 6 November 2019. https://www.bbc.co.uk/news/science-environment-50302392.

40. UNESCO. The United Nations World Water Development Report 2022: Groundwater: Making the Invisible Visible. s.l. : UNESCO, 2022.

41. Chislock, M. F., Doster, E., Zitomer, R. A. & Wilson, A. E. Eutrophication: Causes, Consequences, and Controls in Aquatic Ecosystems. *Nature.com.* [Online] 2013.

https://www.nature.com/scitable/knowledge/library/eutrophication-causes-consequences-and-controls-in-aquatic-102364466/.

42. Eutrophication and recovery in experimental lakes: implications for lake management. Schindler, D W. s.l. : AAAS, 24 May 1974, Science, Vol. 184, pp. 897 - 899.

43. Feature Detail Report For: Gulf of Mexico. *USGS.* [Online] 2000. https://web.archive.org/web/20201231190019/https://geonames.usgs.gov/apex/f?p=GNISPQ:3:::NO::P3_FID:558730.

44. Jamail, Dahr. BP anniversary: Toxicity, suffering and death. *Ajazeera.* [Online] 2011. https://www.aljazeera.com/features/2011/4/16/bp-anniversary-toxicity-suffering-and-death.

45. —. Gulf ecosystem in crisis after BP spill. *Aljazeera.* [Online] 2013. https://www.aljazeera.com/features/2013/10/20/gulf-ecosystem-in-crisis-after-bp-spill/.

46. —. BP dispersants 'causing sickness'. *Aljazeera.* [Online] 2013. https://www.aljazeera.com/features/2010/10/29/bp-dispersants-causing-sickness/.

47. Eutrophication in the Gulf of Mexico: How Midwestern farming practices are creating a 'Dead Zone'. *Dartmouth Undergraduate Journal of Science.* [Online] 2012. https://sites.dartmouth.edu/dujs/2012/03/11/eutrophication-in-the-gulf-of-mexico-how-midwestern-farming-practices-are-creating-a-dead-zone/.

48. *Salad Leaf Juices Enhance Salmonella Growth, Colonization of Fresh Produce, and Virulence.* Giannis Koukkidis, Richard Haigh, Natalie Allcock, Suzanne Jordan, Primrose Freestone. s.l. : National Library of Medicine, 2016.

49. Manure contamination of drinking water influences dairy cattle water intake and preference. Karin E. Schütz, Frances J. Huddart, Neil R. Cox. 2019, Elsevier.

50. Manure-borne pathogens as an important source of water contamination: An update on the dynamics of pathogen survival/transport as well as practical risk mitigation strategies. Oluwadara Oluwaseun Alegbeleye, Anderson S. Sant'Ana. 2020.

51. Agency, Environment. Farmer fined for obstructing Environment Agency officer. 2024.

52. Agency, Environmental. Community service for farmer's polluting dam of manure. 2024.

53. Institute, Food and Agriculture Organization of the United Nations & International Water Management. *More people, more food, worse water? a global review of water pollution from agriculture.* s.l. : FAO, 2018.

54. Lacy, Katherine. The number of U.S. farms continues slow decline. s.l. : USDA, 2025.

55. Diversity, Biological. Environmental Impacts of Extreme Animal Confinement. *Biological Diversity.* [Online] https://www.biologicaldiversity.org/takeextinctionoffyourplate/factory_farms/index.html.

56. Services, Wisconsin Department of Health. Infant Methemoglobinemia (Blue Baby Syndrome). *Wisconsin Department of Health Services.* [Online] 2025. https://www.dhs.wisconsin.gov/water/blue-baby-syndrome.htm.

57. Department for Environment, Food & Rural Affairs and Environment Agency. Using nitrogen fertilisers in nitrate vulnerable zones. *DEFRA.* [Online] 2024. https://www.gov.uk/guidance/using-nitrogen-fertilisers-in-nitrate-vulnerable-zones.

58. Principles of planning and establishment of buffer zones. Correll, David L. s.l. : Elsevier, 2005.

59. Insider, Farming. Legal Incentives for Sustainable Agricultural Practices. *Farming Insider.* [Online] January 2025. https://thefarminginsider.com/legal-incentives-sustainable-agriculture/.

60. *The Blue, Green and Grey Water Footprints of Farm Animals and Animal Products.* M.M. Mekonnen, A.Y. Hoekstra. s.l. : Twente Water Centre, University of Twente, Enschede, The Netherlands, 2010.

61. *A Global Assessment of the Water Footprint of Farm Animal Products.* Mesfin M. Mekonnen, Arjen Y. Hoekstra. s.l. : Springer Nature, 2012.

62. Hannah Ritchie, Pablo Rosado, Max Roser. *Data Page: Share of cereals allocated to animal feed.* s.l. : Our World in Data, 2024.

63. Effect of Mono Cropping on Soil Health and Fertility Management for Sustainable Agriculture Practices: A Review. Tegegn Belete, Eshetu Yadete. s.l. : Research Gate, 2023.

64. Shahbandeh, M. Breakdown of soy production worldwide 2018, by end use. *Statista.* [Online] 2022.
https://www.statista.com/statistics/1254608/soy-production-end-uses-worldwide/.

65. Food, Soya. Soya world production. *Soya Food.* [Online] https://www.soya-food.com/about-soya-food/soya-world-production.html.

66. Soya foods and your health. *The Association of UK Dieticians.* [Online] 2022. https://www.bda.uk.com/resource/soya-foods.html.

67. OECD. OECD-FAO Agricultural Outlook 2024-2033. s.l. : OECD, 2024.

68. Board, Nabraska Corn. Field Corn and Sweet Corn: Corn for Cattle vs Humans. *Nabraska Corn Board.* [Online]
https://nebraskacorn.gov/cornstalk/field-corn-and-sweet-corn-corn-for-cattle-vs-humans/.

69. Maize stomatal responses against the climate change. Serna, Laura. s.l. : Frontiers, 2022.

70. Table, Nutrition. Nutritional Information Corn on the Cob, Cooked. *NutrionTable.com.* [Online]
https://www.nutritiontable.com/nutritions/nutrient/?id=1186.

71. Tunnels, First. The Complete Guide Into the World of Companion Planting. *First Tunnels.* [Online]
https://www.firsttunnels.co.uk/page/Companion-Planting-Guide.

72. Maize/soybean intercropping increases nutrient uptake, crop yield and modifies soil physio-chemical characteristics and enzymatic activities in the subtropical humid region based in Southwest China. Jamal Nasar, Munir Ahmad, Harun Gitari, Li Tang, Yuan Chen, Xun-Bo Zhou. Guangxi Province : BMC Plant Biology, 1 May 2024.

73. Sarfas, Abby Jade. How many animals can you save by going vegan? *The Humane League United Kingdom.* [Online] 18 May 2022.

https://thehumaneleague.org.uk/article/how-many-animals-can-you-save-by-going-vegan.

74. Calculator, Veegan. *vegancalculator.com.* [Online] vegancalculator.com.

75. Society, The Vegan. Definition of Veganism. *The Vegan Society.* [Online] https://www.vegansociety.com/go-vegan/definition-veganism.

76. Data, Food and Agriculture Organization of the United Nations (2025) – with major processing by Our World in. "Number of cattle slaughtered for meat – FAO" [dataset]. *Our World in Data.* [Online] 2025. https://ourworldindata.org/grapher/animals-slaughtered-for-meat.

77. *Intensive poultry farming: A review of the impact on the environment and human health.* Goran Gržinić, Agnieszka Piotrowicz-Cieślak, Agnieszka Klimkowicz-Pawlas, Rafał L. Górny, Anna Ławniczek-Wałczyk, Lidia Piechowicz, Ewa Olkowska, Marta Potrykus, Maciej Tankiewicz, Magdalena Krupka, Grzegorz Siebielec, Lidia Wolska. 3, s.l. : Elsevier, 1 Feb 2023, Science of The Total Environment, Vol. 858.

78. International, Four Paws. The Life Expectancy of Farm Animals. *Four Paws.* [Online] 22 April 2024. https://www.four-paws.org/campaigns-topics/topics/farm-animals/age-of-farm-animals.

79. Viva. Are You a Baby Eater? *Are You a Baby Eaters?* [Poster]. s.l. : Viva, 03 March 2020.

80. Farming, Compassion in World. Chickens farmed for meat. *Compassion in World Farming.* [Online] https://www.ciwf.org.uk/farm-animals/chickens/meat-chickens/.

81. Barckley, Jennifer. Egg Production Facts and Stats: How Eggs End Up On Our Plates. *The Humane League.* [Online] 30 MArch 2021. https://thehumaneleague.org/article/egg-production.

82. Affairs, Department for Environment and Rural. Code of practice for the welfare of meat chickens and meat breeding chickens. s.l. : GOV.UK, 25 January 2024.

83. UNION, THE COUNCIL OF THE EUROPEAN. Laying down minimum standards for the protection of laying hens. *Council Directive 1999/74/EC of 19 July 1999 laying down minimum standards for the*

protection of laying hens. s.l. : THE COUNCIL OF THE EUROPEAN UNION, 19 July 1999.

84. England, Campaign to Protect Rural. CPRE's Vision for the future of farming:. *CPRE.* [Online] https://www.cpre.org.uk/wp-content/uploads/2019/11/The_future_of_pig_and_poultry_farming.pdf.

85. *Sustainable poultry farming practices: a critical review of current strategies and future prospects.* Ramesh Bahadur Bist, Keshav Bist, Sandesh Poudel, Deepak Subedi, Xiao Yang, Bidur Paneru, Sudhagar Mani, Dongyi Wang, Lilong Chai. 12, December 2024, Poultry Science, Vol. 103. 104295.

86. Monnier, A. A. What is responsible antibiotic use? A search for a multi-stakeholder definition, quality. Hasselt : Radboud Respiritory, 2021.

87. Characterization of commercial poultry farms in Mexico: Towards a better understanding of biosecurity practices and antibiotic usage patterns. Erika Ornelas-Eusebio, Gary García-Espinosa, Karine Laroucau, Gina Zanella. 12, s.l. : PloS one, 1 December 2020, Vol. 15.

88. *How did antibiotic growth promoters increase growth and feed efficiency in poultry?* Mariano Enrique Fernández Miyakawa, Natalia Andrea Casanova, Michael H Kogut. 2, s.l. : Elsevier, 17 November 2023, Poultry Science, Vol. 103.

89. *Pollution by Antibiotics and Antimicrobial Resistance in LiveStock and Poultry Manure in China, and Countermeasures.* Ming Tian, Xinmiao He, Yanzhong Feng, Wentao Wang, Heshu Chen, Mong Gong, Di Liu, Jihong Liu Clarke, André van Eerde. 5, Basel, Switzerland : s.n., 6 May 2021, Antibiotics, Vol. 10.

90. Tabler, G. Tom. Water Intake: A Good Measure of Broiler Performance. *The Poultry Site.* [Online] 17 December 2003. https://www.thepoultrysite.com/articles/water-intake-a-good-measure-of-broiler-performance.

91. Henhouse, Kropper Mobile. What is the water requirement of laying hens? *Kropper Mobile Henhouse.* [Online] 27 January 2023. https://mobilekropper.com/what-is-the-water-requirement-of-laying-hens/.

92. Wills, Ruth. A step-by-step guide to cleaning your poultry shed. *Farmers Weekly.* [Online] 7 November 2020.

https://www.fwi.co.uk/livestock/poultry/a-step-by-step-guide-to-cleaning-your-poultry-shed.

93. LIVE BIRD TRANSPORT AND RECEPTION. *Poultry Processing Equipment.* [Online] https://www.poultryprocessingequipment.com/live_bird_transport_and_reception/live_bird_crate_washer.asp.

94. RBK3 rotating scalding tank for poultry. *Burdis Poultry.* [Online] https://www.burdis-poultry.com/poultry-scalding/54-rbk3-rotating-scalding-tank-for-poultry.html.

95. Justice, United States Department of. *Tyson Foods Pleads Guilty and Agrees to Pay Fine for OSHA Violation That Led to Worker Death.* United States Department of Justice, U.S. District Court in Arkansas. s.l. : USDOJ, 2009.

96. Technology, Massachusetts Institute of. What makes methane a more potent greenhouse gas than carbon dioxide? *MIT Climate Portal.* [Online] 7 December 2023. https://climate.mit.edu/ask-mit/what-makes-methane-more-potent-greenhouse-gas-carbon-dioxide.

97. Technology, Massachusetts Institue of. Fertilizer and Climate Change. *MIT Climate Portal.* [Online] 15 July 2021. https://climate.mit.edu/explainers/fertilizer-and-climate-change.

98. *Poultry litter ash characterisation and recovery.* Fahimi A, Bilo F, Assi A, Dalipi R, Federici S, Guedes A, Valentim B, Olgun H, Ye G, Bialecka B, Fiameni L, Borgese L, Cathelineau M, Boiron MC, Predeanu G, Bontempi E. s.l. : National Institute of Medicine, 15 June 2020.

99. *The burden of cancer attributable to dietary dioxins and dioxin-like compounds exposure in China, 2000–2020.* Ziwei Shi, Yiling Li, Xiaohan Song, Yibaina Wang, Jianwen Li, Sheng Wei. s.l. : Elsevier, December 2024, Environment International, Vol. 194.

100. Sarfas, Abby Jade. How farmers increase egg production in hens. *The HUmane League United Kingdom.* [Online] 8 October 2021. https://thehumaneleague.org.uk/article/how-farmers-increase-egg-production-in-hens.

101. What happens to male chicks in the egg industry? *The Humane League United Kingdom.* [Online] 21 March 2024.

https://thehumaneleague.org.uk/article/what-happens-to-male-chicks-in-the-egg-industry.

102. Millstein, Seth. Why Is Chick Culling Still a Thing When Other Technologies Exist? *Sentient.* [Online] 10 June 2024. https://sentientmedia.org/why-chick-culling/.

103. Cages, Open. *The Price of British Chicken: How Supermarkets Are.* s.l. : Opencages.org, 2022. pp. 2 - 3.

104. *Particulate matter in poultry house on poultry respiratory disease: a systematic review.* Kai Wang, Dan Shen, Pengyuan Dai, Chunmei Li. 4, April 2023, Poultry Science, Vol. 102.

105. Cobb-Vantress. The Cobb Broiler Management Guide.

106. *Environmental Exposure to Confined Animal Feeding Operations and Respiratory Health of Neighboring Residents.* Katja Radon, Anja Schulze, Vera Ehrenstein, Rob T. van Strien, Gerog Praml, Nowak, Dennis. 3, May 2007, Epidemiology, Vol. 18.

107. *Evaluating draft Environmental Protection Agency emissions models for broiler operations.* G. Li, R.S. Gates, Y. Xiong, b. c. Ramirez, R. T. Burns. 4, December 2023, Journal of Applied Poultry Research, Vol. 32.

108. *The prevalence and concentration of Salmonella enterica in poultry litter in the southern United States.* Laurel L Dunn, Vijendra Sharma, Travis K Chapin, Loretta M Friedrich, Colleen C Larson, Camila Rodrigues, Michele Jay-Russell, Keith R Schneider, Michelle D Danyluk. s.l. : PLoS One, 26 May 2022.

109. *Current state of poultry waste management practices in Bangladesh, environmental concerns, and future recommendations.* Md Masudur Rahman, Alamgir Hasan, Ismail Hossain. 3, September 2022, Journal of Advanced Veterinary and Animal Research , Vol. 9.

110. Maryland, District Court of. Waterkeeper Alliance, Inc v Alan & Kristin Hudson Farm et al. s.l., Maryland, United States : District Court of Maryland, 2013.

111. *Bacterial diversity changes in agricultural soils influenced by poultry litter fertilization.* Parente, C.E.T., Brito, E.M.S., Caretta, C.A. et al. 15 February 2021, Braz J Microbiol, Vol. 52, pp. 675–686.

112. *Increased risk of pneumonia in residents living near poultry farms: does the upper respiratory tract microbiota play a role.* Floor Borlée, C Joris Yzermans, Bernadette Aalders, Jos Rooijackers, Esmeralda Krop, Catharina B M Maassen, François Schellevis, Bert Brunekreef, Dick Heederik, Lidwien A M Smit. 9, 1 November 2021, Am J Respir Crit Care Med, Vol. 196, pp. 1152-1161.

113. Human milk antibiotic residue levels and their relationship with delivery mode, maternal antibiotic use and maternal dietary habits. M Dinleyici, G K Yildirim, O Aydemir, T B Kaya, Y Bildirici, K B Carman. 19, October 2018, Eur Rev Med Pharmacol Sci, Vol. 22, pp. 6560-6566.

114. Texas, District Court of. Steve Huynh, Individually, Yvonne Huynh, Individually Huynh Poultry Farm, LLC d/b/a Steve Thi Huynh Poultry Farm d/b/a Huynh Poultry Farm, T & N Poultry Farm, LLC, Thinh Bao Nguyen, Individually, Timmy Huynh Poultry Farm, Timmy Huynh, Individually and San. 2021.

115. Project, Environmental Intergrity. Poultry Industry Ammonia Air Pollution Adds More Nitrogen to Chesapeake Bay than all MD or PA Sewage Plants. EIP, EIP. 2020.

116. Parson, Will. *Agricultural Runoff.* Chesapeake Bay Program. s.l. : CheasapeakeBay.net.

117. Wheeler, Timothy B. Maryland Orders Chicken Rendering Plant Shut Down, and Update. *Bay Journal.* [Online] 4 January 2022. https://www.commonsenseeasternshore.org/maryland-orders-chicken-rendering-plant-shut-down-and-update.

118. Service, National Agricultural Statistics. *Agricultural Statistics 2020.* s.l. : National Agricultural Statistics Service, 2020.

119. Commission, European. Nitrates. *European Commission.* [Online] https://environment.ec.europa.eu/topics/water/nitrates_en.

120. Agency, European Environment. The Natura 2000 protected areas network. *European Environment Agency.* [Online] 28 Feburary 2023. https://www.eea.europa.eu/themes/biodiversity/natura-2000.

121. Ralte, Rachel. More Fipronil-tainted Eggs Found in Korea; Govt Apologises for Fiasco. *The Poultry Site.* [Online] 18 August 2017.

https://www.thepoultrysite.com/news/2017/08/more-fiproniltainted-eggs-found-in-korea-govt-apologises-for-fiasco.

122. Choi, Juho. South Korea's Struggle Against Avian Influenza. *KEI.* [Online] 12 January 2017. https://keia.org/the-peninsula/south-koreas-struggle-against-avian-influenza/.

123. England, The Environment Agency and Natural. Salmon stocks in England lowest on record. *GOV.UK.* [Online] 7 10 2024. https://www.gov.uk/government/news/salmon-stocks-in-england-lowest-on-record.

124. FAO. The State of World Fisheries and Aquaculture 2022. Towards Blue Transformation. FAO. Rome : s.n., 2022.

125. Foundation, Environmental Justice. *What's the Ctach.* Environmental Justice Foundation. 2005.

126. Henry, Leigh. Bycatch. *WWF.* [Online] https://www.worldwildlife.org/threats/bycatch.

127. Taylor, Barbara. Clear Signs of Progress in Protecting Endangered Vaquita. *Sea Shepherd.* [Online] 7 June 2023. https://seashepherd.org/2023/06/07/clear-signs-of-progress-in-protecting-endangered-vaquita/.

128. Shailer, Daniel. Fishers decry 'underhanded' new initiative to protect Mexico's vaquita. *Mongabay.* [Online] 24 May 2024. https://news.mongabay.com/2024/05/fishers-decry-underhanded-new-initiative-to-protect-mexicos-vaquita/.

129. Association, The Marine Biological. Protecting marine wildlife: New EU project aims to reduce bycatch in fishing. *The Marine Biological Association.* [Online] 12 December 2023. https://www.mba.ac.uk/protecting-marine-wildlife-new-eu-project-aims-to-reduce-bycatch-in-fishing/.

130. NOAA. Underwater noise de creases whale communications in Stellwagen Bank Sanctuary. *National Oceanic and Atmospheric Administration.* [Online] 15 August 2012. www.sciencedaily.com/releases/2012/08/120515142050.htm.

131. Banna, Karim. Destructive Fishing Practices: How They Affect Coral Reefs and Their Resilience. *Fisging and Fish.* [Online] 17 11 2024. https://fishingandfish.com/how-do-destructive-fishing-practices-affect-the-coral-reef/.

132. Hermann, EMily. Innovation in river dolphin conservation. *WWF.* [Online] 7 December 2021. https://www.worldwildlife.org/stories/innovation-in-river-dolphin-conservation.

133. Ritchie, Hannah. Most plastic in the Great Pacific Garbage Patch comes from the fishing industry. *OurWorldinData.org.* [Online] 2023. https://ourworldindata.org/plastic-great-pacific-garbage.

134. Microplastics in different tissues of a pelagic squid (Dosidicus gigas) in the northen Humboldt Current Ecosystem. Y Gong, Y Wang, L Chen, Y Li, X Chen, B Liu. s.l. : Mar Pollut Bull, 21 May 2021.

135. Mike Merritt, Rob Edwards. The madness of filleting Scottish fish in China. *Robedwards.com.* [Online] 23 August 2009. https://www.robedwards.com/2009/08/the-madness-of-filleting-scottish-fish-in-china.html.

136. Data, Food and Agriculture Organization of the United Nations (via World Bank) (2025) – processed by Our World in. Aquaculture. s.l. : FAO, 2025.

137. FAO. 4. AQUACULTURE METHODS AND PRACTICES: A SELECTED REVIEW. FAO. s.l. : FAO.

138. Staff. RAS Fish Farming: Start Recirculating Aquaculture System. *Roys Farm.* [Online] 26 August 2024. https://www.roysfarm.com/ras-fish-farming/.

139. Edwards, Peter. Aquaculture environment interactions: Past, present and likely future trends. *Aquaculture.* 1 October 2015, Vol. 447, pp. 2-14.

140. Foundation, FISH Safety. Triggering Death: Quantifying the true human cost of global fishing. *FISH Safety Foundation.* [Online] 2022. https://fishsafety.org.projects.the-human-cost-of-fishing/.

141. McVeigh, Karen. Fish From Farms Threaten Wild Atlantic Salmon Population. *The Guardian.* [Online] 30 September 2023.

https://www.theguardian.com/environment/2023/sep/30/thousands-of-salmon-escaped-an-icelandic-fish-farm-the-impact-could-be-deadly.

142. Roy P. E. Yanong, Ruth Francis-Floyd, Barbara D. Petty. Parasitic Diseases in Aquaculture. *MSD Manual Veterinary Manual.* [Online] October 2021. https://www.msdvetmanual.com/exotic-and-laboratory-animals/aquaculture/parasitic-diseases-in-aquaculture.

143. —. Bacterial Diseases in Aquaculture. *MSD Manual Veterinary Manual.* [Online] October 2021. https://www.msdvetmanual.com/exotic-and-laboratory-animals/aquaculture/bacterial-diseases-in-aquaculture.

144. —. Viral Diseases in Aquaculture. *MSD Manual Veterinary Manual.* [Online] October 2021. https://www.msdvetmanual.com/exotic-and-laboratory-animals/aquaculture/viral-diseases-in-aquaculture.

145. Sandvik, Kjersti. Pollution in Open Net Pen Salmon Farming. *NASF.* [Online] https://nasf.is/en/pollution/.

146. Appel, Loreto. Chilean salmonid exports were worth a record $6.6 bn last year. *Fish Farming Expert.* [Online] 13 January 2023. https://www.fishfarmingexpert.com/chile/chilean-salmonid-exports-were-worth-a-record-66-bn-last-year/1475940.

147. Wendy Norden, Rolando Ibarra, Daniela Farias, Maria Lorena Gonzales. Farmed salmon in Chile. *Seafood Watch.* [Online] https://www.seafoodwatch.org/our-projects/farmed-salmon-in-chile.

148. Augustinis, Francesco De. Natural paradise in Chile under threat from fish farming. *The Ferret.* [Online] 15 February 2023. https://theferret.scot/nature-chile-under-threat-from-fish-farming/.

149. *1 - The Concept of Stress in Fish.* Carl B. Schreck, Lluis Tort. 2016, Fish Physiology, Vol. 25, pp. 1-34.

150. Auchterlonie, Dr. Neil. Aquaculture's input efficiency shines as FIFO ratios improve. *Global Seafood Alliance.* [Online] 23 October 2017. https://www.globalseafood.org/advocate/aquaculture-input-efficiency-fifo/.

151. Mittal, Renu. Overfishing Threatens Chile's Small-Scale Fisheries. *Walton Familiy Foundation.* [Online] 8 November 2017. [Cited: 28 October 2024.]

https://www.waltonfamilyfoundation.org/stories/environment/overfishing-threatens-chiles-small-scale-fisheries.

152. *Conservation aquaculture: Shifting the narrative and paradigm of aquaculture's role in resource management.* Halley E. Froehlich, Rebecca R. Gentry, Benjamin S. Halpern. November 2017, Biological Conservation, Vol. 215, pp. 162-168.

153. Times, Tasmanian. Big Stink Over Mass Salmon Mortality. *TasmanianTimes.com.* [Online] 23 october 2024. https://tasmaniantimes.com/2024/10/big-stink-over-mass-salmon-mortality/.

154. —. StatementsLong Bay Protest against Tassal Salmon Pens. *TasmanianTimes.com.* [Online] 27 August 2022. https://tasmaniantimes.com/2022/08/long-bay-protest-against-tassal-salmon-pens/.

155. Anti-salmon farm protestors take to their boats in the D'Entrecasteaux channel. news.com.au; YouTube, 2023.

156. Alliance, Social Reform. Social License to Operate. *Social Reform Alliance.* [Online] 2022. https://www.salmonreform.org/social-license.

157. Shiven Batra, Daniel Palmer. *A Deep Dive into the CAFO Industry's Impact and Evolution.* Institute for Youth in Policy. 2025.

158. *Increased transparency in accounting conventions could benefit.* Wedderburn-Bisshop, Gerard. s.l. : IOP Publishing, 11 March 2025, Environmental Research Letters.

159. *Climate change and livestock: Impacts, adaptation, and mitigation.* M. Melissa Rojas-Downing, A. Pouyan Nejadhashemi, Timothy Harrigan, Sean A. Woznicki. 2017, Climate Risk Management, Vol. 16, pp. 145-163.

160. Cambria Glosz, Jenny Hughes. Why Is Grass-Fed Beef Better? *Nutrition Insider.* [Online] 31 October 2024. https://thenutritioninsider.com/wellness/why-is-grass-fed-beef-better/.

161. *Grass-fed vs. grain-fed beef systems: performance, economic, and environmental trade-offs.* Sarah C Klopatek, Elias Marvinney, Toni Duarte, Alissa Kendall, Xiang (Crystal) Yang, James W Oltjen. 2, 21 December 2021, Journal of Animal Science, Vol. 100.

162. A global meta-analysis of livestock grazing impacts on soil properties. Liming Lai, Sandeep Kumar. s.l. : PLoS ONE, 7 August 2020.

163. Department of Agriculture, Environment and Rural Affairs. Beef farming technology projects.

164. Cornelis De haan, Henning Steinfeld, Harvey D. Blackburn, Commission of the European Communities, World Bank. *Livestock & the environment : finding a balance.* s.l. : SERBIULA (sistema Librum 2.0).

165. *The water footprint of poultry, pork and beef: A comparative study in different countries and production systems.* Winnie Gerbens-Leenes, Mesfin Mergia Mekonnen, Arjen Hoekstra. 2, s.l. : Elsevier, March 2013, Water Resources and Industry, Vol. 1, pp. 25-36.

166. Brodie, Herbert. Estimating Irrigation Water Requirements. *Irrigation Toolbox.* [Online] https://www.irrigationtoolbox.com/ReferenceDocuments/Extension/Eastern%20States/ESTIMATING%20IRRIGATION%20WATER%20REQUIREMENTS.pdf.

167. Rasby, Rick. Determining How Much Forage a Beef Cow Consumes Each Day. *University of Nebraska–Lincoln.* [Online] 2 April 2020. https://beef.unl.edu/cattleproduction/forageconsumed-day/.

168. Miller, Rhonda. How Much Manure Will My Animals Produce? *Extension Utah State University.* [Online] https://extension.usu.edu/smallfarms/files/How_Much_Manure.pdf.

169. Tim G. Benton, Carling Bieg, Helen Harwatt, Roshan Pudasaini, Laura Wellesley. *Food system impacts.* Energy, Environment and Resources Programme, Chatham House. s.l. : Chatham House, 2021.

170. *Declining biodiversity for food and agriculture needs urgent global action.* Dafydd Pilling, Julie Bélanger, Irene Hoffmann. 3, February 2020, Nature Food, Vol. 1.

171. Quinton, Amy. Meet Cosmo, a Bull Calf Designed to Produce 75% Male Offspring. *UC Davis.* [Online] 23 July 2020. https://www.ucdavis.edu/food/news/placeholder-cow-story.

172. Furness/Undark, Dyllan. Gene-edited cows could make meat more sustainable. But would people eat it? *Popular Science.* [Online] 7 August

2020. https://www.popsci.com/story/technology/gene-edited-cows-benefits/.

173. Reuters. Explainer: How four big companies control the U.S. beef industry. *Reuters.* [Online] 17 June 2021. https://www.reuters.com/business/how-four-big-companies-control-us-beef-industry-2021-06-17/.

174. Bentlage, Darvin. Just four meatpackers control 85% of the market. Cattlemen like me need a voice. *Minnesota Reformer.* [Online] 14 November 2022. https://minnesotareformer.com/2022/11/14/just-four-meatpackers-control-85-of-the-market-cattlemen-like-me-need-a-voice/.

175. Department of Agriculture, Environment and Rural Affairs. Beef management. *Department of Agriculture, Environment and Rural Affairs.* [Online] https://www.daera-ni.gov.uk/articles/beef-management.

176. Extension, University of Georgia. Lawton Stewart. *University of Georgia Extension.* [Online] 2024. https://extension.uga.edu/publications/detail.html?number=B895&title=mineral-supplements-for-beef-cattle.

177. Protocol, Greenhouse Gas. *IPCC Global Warming Potential Values.* s.l. : Greenhouse Gas Protocol, 2024.

178. R.K. Pachauri, L.A. Meyer. IPCC, 2014: Climate Change 2014: Synthesis Report. Contribution of Working Groups I, II and III to the Fifth Assessment Report of the Intergovernmental Panel on Climate Change. IPCC. s.l. : IPCC, 2014. p. 151.

179. UNFCCC. GHG data from UNFCCC. s.l. : UNFCCC.

180. Data and statistics. *Data and statistics.* s.l. : IEA.

181. Data, Our World in. Emissons. s.l. : OWID.

182. Institute, World Resource. s.l. : World Resource Institute.

183. IATA. IATA Per Passenger CO2 Emissions Calculator: Approach and Documentation. s.l. : IATA.

184. ICCT. Charts and Visualizations. s.l. : ICCT.

185. FAO. *Methane emissions in livestock and rice systems.* FAO. Rome : FAO, 2023. Book (stand alone).

186. EPA. Greenhouse Gas Emissions from a Typical Passenger Vehicle. EPA. s.l. : EPA.

187. Department of Agriculture, Environment and Rural Affairs. *What is the Beef Carbon Reduction (BCR) Scheme.* Department of Agriculture, Environment and Rural Affairs. s.l. : Department of Agriculture, Environment and Rural Affairs.

188. Hilary Osborne, Bibi van der Zee. Live export: animals at risk in giant global industry. *The Guardian.* [Online] 20 Janurary 2020. https://www.theguardian.com/environment/2020/jan/20/live-export-animals-at-risk-as-giant-global-industry-goes-unchecked.

189. Agriculture, Australian Government Department of. *Mortality Investigation Report 53.* Department of Agriculture, Australian Government. s.l. : Department of Agriculture, 2015. p. 6.

190. Brad Lendon, Emiko Jozuka. Japan Coast Guard finds third sailor as search for missing ship with 43 sailors and 5,800 cows aboard continues. *CNN.* [Online] 4 September 2020. https://edition.cnn.com/2020/09/03/asia/typhoon-maysak-cargo-ship-missing-japan-intl-hnk-scli/index.html.

191. Ritchie, Hannah. *How many animals are factory-farmed?* OWID. s.l. : OWID, 2023.

192. Gateway, Pork Information. How To Process Piglets. *Pork Information Gateway.* [Online] https://porkgateway.org/resource/how-to-process-piglets/.

193. Ray Massey, Ann Ulmer. *Pork Production and Greenhouse Gas Emissions.* Swine Extension. s.l. : swineextension.org, 2019.

194. C. S. Dunkley, K. D. Dunkley. *Greenhouse Gas Emissions from Livestock and Poultry.* s.l. : Agriculture, Food and Analytical Bacteriology, 2013. p. 13.

195. Charlie Hope-D'Anieri, Austin Frerick,. The hog baron. *Food and Environment Reporting Network.* [Online] 19 August 2021. https://thefern.org/2021/04/the-hog-baron/.

196. Equality, Animal. Investigation: Pollution from Mexico's factory farming exposed. *Animal Equality.* [Online] 6 August 2020.

https://animalequality.org/news/2020/08/06/investigation-animal-equality-exposes-environmental-pollution-from-mexicos-industrial-farming/.

197. Marty Matlock, Greg Thoma, Eric Boles, Mansoor Leh, Heather Sandefur, Rusty Bautista, Rick Ulrich. *A Life Cycle Analysis of Water Use in.* s.l. : Pork Checkoff. p. 76.

198. James J. Hoorman, Jonathan N. Rausch, Larry C. Brown. Guidelines for Applying Liquid Animal Manure to Cropland with Subsurface and Surface Drains. *Ohioline.* [Online] 29 October 2009. [Cited: 15 12 2024.] https://ohioline.osu.edu/factsheet/ANR-21.

199. NCpedia. The Impact of Hog Farms. *Anchor.* [Online] 1995. https://www.ncpedia.org/anchor/impact-hog-farms.

200. Department of Agriculture, Environment and Rural Affairs. Water advice for livestock farmers. *The Department of Agriculture, Environment and Rual Affairs.* [Online] https://www.daera-ni.gov.uk/articles/water-advice-livestock-farmers.

201. al., Alexander et. Meat conversion efficiencies. *Our World in Data.* [Online] 2016. processed by Our World in Data. https://ourworldindata.org/grapher/feed-required-to-produce-one-kilogram-of-meat-or-dairy-product.

202. Poultry, WATT. Cleaning questions. *WATT Poultry.* [Online] 1 July 2019. https://www.wattagnet.com/home/article/15475427/cleaning-questions-cleaning-pig-pens.

203. Kate. The science of slaughter: Is the stunning of pigs in UK slaughterhouses currently humane? *On Animals.* [Online] 6 May 2022. https://onanimals.co.uk/2022/05/06/science-of-slaughter-is-stunning-of-pigs-humane/.

204. Cade, DL. Nat Geo Photographer George Steinmetz Arrested for Taking Photos of Feedlot. *PetaPixel.* [Online] 11 July 2013. https://petapixel.com/2013/07/11/nat-geo-photographer-george-steinmetz-arrested-for-taking-photos-of-feedlot/.

205. *Defecation frequency and timing, and stool form in the general population: a prospective study.* K W Heaton, J Radvan, H Cripps, R A Mountford, F E Braddon, A O Hughes. 6, 1992, Gut, Vol. 3, pp. 818-824.

206. *Characterisation of pig manure for methane emission modelling in Sub-Saharan Africa.* Ngwa M. Ngwabie, Bren N. Chungong, Fabrice L. Yengong. June 2018, Biosystems Engineering, Vol. 170, pp. Pages 31-38.

207. Moon, Emily. North Carolina's Hog Waste Problem Has a Long History. Why Wasn't It Solved in Time for Hurricane Florence? *Pacific Standard.* [Online] 2018. https://psmag.com/environment/why-wasnt-north-carolinas-hog-waste-problem-solved-before-hurricane-florence/.

208. Tony Dutzik, Piper Crowell, John Rumpler. *Wasting Our Waterways Toxic Industrial Pollution and the Unfulfilled Promise of the Clean Water Act.* Frontier Group, Environment America Research & Policy Center. s.l. : Environment America Research and Policy Center, 2009.

209. EWG. Pouring It On Nitrate Contamination of Drinking Water. s.l. : EWG, 1996.

210. Mirvish, S S. The etiology of gastric cancer. Intragastric nitrosamide formation and other theories. *J Natl Cancer Inst.* 1983, Vol. 71, 3, pp. 629-647.

211. —. The significance for human health of nitrate, nitrite, and n-nitroso compounds. in Nitrate Contamination: Exposure, Consequence, and Control. 1991.

212. R Gray, R Peto, P Brantom, P Grasso. Chronic nitrosamine ingestion in 1040 rodents: the effect of the choice of nitrosamine, the species studied, and the age of starting exposure. *Cancer Res.* 1991, Vol. 51, 23 Pt 2, pp. 6470-6491.

213. C Cuello, P Correa, W Haenszel, G Gordillo, C Brown, M Archer, S Tannenbaum. Gastric cancer in Colombia. I. Cancer risk and suspect environmental agents. *J Natl Cancer Inst.* 1976, Vol. 57, 5, pp. 1015-1020.

214. van Maanen, J. M. S., van Dijk, A., Mulder, K., de Baets, M. H., Menheere, P. C. A., van der Heide, D., Mertens, P. L. J. M., & Kleinjans, J. C. S. Consumption of drinking water with high nitrate levels causes hypertrophy of the thyroid. *Toxicology Letters.* 1994, Vol. 72, pp. 365-374.

215. M M Dorsch, R K Scragg, A J McMichael, P A Baghurst, K F Dyer. Congenital malformations and maternal drinking water supply in rural South Australia: a case-control study. *Am J Epidemiol.* 1984, Vol. 119, 4, pp. 473-486.

216. Knox, E G. Anencephalus and Dietary Intakes. *Journal of Preventive and Social Medicine.* 1972, Vol. 26, 4, pp. 219–223.

217. M Super, H de V. Heese, D. MacKenzie, W.S. Dempster, J. du Plessis, J.J. Ferreira. An epidemiological study of well-water nitrates in a group of south west african/namibian infants. *Water Research.* 1981, Vol. 15, 11, pp. 1265-1270.

218. Graff, Michael. Millions of dead chickens and pigs found in hurricane floods. *The Guardian.* [Online] 22 September 2018. https://www.theguardian.com/environment/2018/sep/21/hurricane-florence-flooding-north-carolina?blm_aid=37287.

219. Jim Efstathiou Jr., Sylvia Carignan, Shruti Date Singh. Toxic Waste. Animal Manure. Hurricane Florence Could Be a Public Health Disaster. *Time.* [Online] 18 September 2018. https://time.com/5392478/hurricane-florence-risks-sludge-manure/.

220. Shahbandeh, M. Top U.S. states by number of hogs and pigs 2024. *Statista.* [Online] 30 July 2024. https://www.statista.com/statistics/194371/top-10-us-states-by-number-of-hogs-and-pigs/.

221. Kealey, Kate. Happy National Pig Day! Here are five fun facts about pigs in Iowa. *Des Moines Register.* [Online] 01 March 2024. [Cited: 01 March 2025.] https://eu.desmoinesregister.com/story/entertainment/2024/03/01/five-fun-facts-about-pigs-in-iowa-on-national-pig-day/72804722007/.

222. Kleese, Nick. Out of Iowa. *Climate Generation.* [Online] 31 October 2024. https://climategen.org/blog/out-of-iowa/.

223. USDA. *2022 Census of Agriculture.* US Department of Agriculture. 2024.

224. USDA/NASS. 2024 STATE AGRICULTURE OVERVIEW Iowa. s.l. : USDA, 2024.

225. Group, Iowa State University and The University of Iowa Study. *Iowa Concentrated Animal Feeding Operations.* Iowa State University. 2022. Study.

226. Watch, Food and Water. *In Hot Water: Iowa's Animal.* 2024. p. 3.

227. Terry, Jennifer. NEWS RELEASE: Des Moines Water Works begins operation of Nitrate Removal Facility because of nutrient spikes in raw source water. *Des Moines Water Works.* [Online] 09 January 2022. https://www.dmww.com/news_detail_T37_R328.php.

228. *A review of new emerging livestock-associated methicillin-resistant Staphylococcus aureus from pig farms.* Aswin Rafif Khairullah, Shendy Canadya Kurniawan, Mustofa Helmi Effendi, Sri Agus Sudjarwo, Sancaka Chasyer Ramandinianto, Agus Widodo, Katty Hendriana Priscilia Riwu, Otto Sahat Martua Silaen, Saifur Rehman. 1, 2023, Veterinary World, Vol. 16, pp. 46-58.

229. A Systematic Review Analyzing the Prevalence and Circulation of Influenza Viruses in Swine Population Worldwide. Ravendra P Chauhan, Michelle L Gordon. 08 May 2020.

230. Goldstein, Bennett. The chairman: How a plan to develop Wisconsin's largest pig farm upended a small town's politics. *Croix 360.* [Online] 15 December 2023. https://www.stcroix360.com/2023/12/the-chairman-how-a-plan-to-develop-wisconsins-largest-pig-farm-upended-a-small-towns-politics/.

231. *Interactions between soil structure dynamics, hydrological processes, and organic matter cycling: A new soil-crop model.* Nicholas Jarvis, Elsa Coucheney, Elisabet Lewan, Katharina H. E. Meurer, Thomas Keller, Mats Larsbo. 2, 10 March 2024, European Journal of Soil Science, Vol. 75.

232. Society, The Royal. *Soil structure.* The Royal Society. s.l. : The Royal Society, 2020.

233. Arbor J.L. Quist, David A. Holcomb, Mike Dolan Fliss, Paul L. Delamater, David B. Richardson, Lawrence S. Engel. Exposure to industrial hog operations and gastrointestinal illness in North Carolina, USA. *Science of The Total Environment.* 15 July 2022, Vol. 830.

234. P. Derek Petersen, Meredith Weinberg, Christopher S. Coleman, Katherine E. May. Large Jury Verdicts in Hog Nuisance Cases Signal CAFO Litigation Is Rising. *Perkins Coie.* [Online] 09 August 2018. https://perkinscoie.com/insights/update/large-jury-verdicts-hog-nuisance-cases-signal-cafo-litigation-rising.

235. Sikowis (Christine Nobiss), Plains Cree. End-Stage Iowa: Big-Ag's Sacrifice Zone and Indigenous Resistance. *Great Plains Action Society.*

[Online] March 2021. https://www.greatplainsaction.org/single-post/end-stage-iowa.

236. *Producing industrial pigs in southwestern China: The rise of contract farming as a coevolutionary process.* Qian Forrest Zhang, Hongping Zeng. 1, s.l. : Wiley Online, 27 October 2021, Journal of Agrarian Change, Vol. 22, pp. 97-117.

237. USDA. *2020 California Almond Forecast.* California Department of Food and Agriculture. 2020. p. 2.

238. Prevalence of colonization by methicillin-resistant Staphylococcus aureus ST398 in pigs and pig farm workers in an area of Catalonia, Spain. Esteban Reynaga, Marian Navarro, Anna Vilamala, Pere Roure, Manuel Quintana, Marian Garcia-Nuñez, Raül Figueras, Carmen Torres, Gianni Lucchetti, Miquel Sabrià. 28 Novermber 2016, BMC Infect Dis, Vol. 16.

239. Associations of Metrics of Peak Inhalation Exposure and Skin Exposure Indices With Beryllium Sensitization at a Beryllium Manufacturing Facility. M Abbas Virji, Christine R Schuler , Jean Cox-Ganser , Marcia L Stanton, Michael S Kent , Kathleen Kreiss , Aleksandr B Stefaniak. 8, 9 September 2019, Annals of Work Exposures and Health, Vol. 63, pp. 856-869.

240. Agriculture, U.S. Department of. USDA issues studies on injury risk to poultry, swine line workers. *Agri-Pulse Comminications, nc.* 2025.

241. Administration, Occupational Safety and Health. Statement from Acting Labor Secretary on USDA study's worker safety data. 13 January 2025.

242. *Slaughtering for a living: A hermeneutic phenomenological perspective on the well-being of slaughterhouse employees.* Karen Victor, Antoni Barnard. 1, 23 February 2016, International Journal of Qualitative Studies on Health and Well-being, Vol. 11. 30266.

243. Cancer, Internation Agency for Research on. IARC Monographs on the Evaluation of Carcinogenic Risks to Humans: Volume 114. Red and Processed Meat. s.l. : IARC, 2018. 978-92-832-0152-6.

244. News, Duke Health. N.C. Residents Living Near Large Hog Farms Have Elevated Disease, Death Risks. *Duke Surgery.* [Online] 2018.

https://surgery.duke.edu/news/nc-residents-living-near-large-hog-farms-have-elevated-disease-death-risks.

245. *Prevalence of Livestock-Associated MRSA in.* Brigitte A. van Cleef, Erwin J. M. Verkade, Mireille W. Wulf, Anton G. Buiting, Andreas Voss, Xander W. Huijsdens, Wilfrid van Pelt, Mick N. Mulders, Jan A. Kluytmans. s.l. : PLoS One, 25 February 2010.

246. Theodore M. Bailey, Peter M. Schantz,. *Trichinosis Surveillance, United States, 1986.* Parasitic Diseases Branch Division, Parasitic Diseases Center for Infectious Diseases. s.l. : CDC, 1988.

247. Intergrity, Environmental. *Raising a Stink: Air Emissions from Factory Farms.* s.l. : Environmental Integrity.

248. A retrospective on claims regarding clinical mastitis in the subsequent lactation, after use of an internal teat sealant in the dry period. Hillerton JE, Berry EA. 2, 17 May 2022, Journal of Dairy Research, Vol. 89, pp. 173-177.

249. Farming, Compassion in World. Calves reared for veal. *Compassion in World Farming.* [Online] https://www.ciwf.org.uk/farm-animals/cows/veal-calves/.

250. Project, Farm Transparency. Age of animals slaughtered. *Farm Transparency Project.* [Online] 18 June 2024. https://www.farmtransparency.org/kb/food/abattoirs/age-animals-slaughtered.

251. morgan-swinhoe, sophia. Male animals in the dairy industry. *Dyfi Dairy.* [Online] https://www.dyfidairy.com/musings-and-articles/male-animals-dairy-industry.

252. Dairying, British. Sexed semen… 20 years later. *British Dairying.* [Online] 16 February 2024. https://www.britishdairying.co.uk/2024/02/16/sexed-semen-20-years-later/.

253. Board, Agriculture and Horticulture Development. Fedding milk and milk replacers to calves. *AHDB.* [Online] https://ahdb.org.uk/Contents/Item/Display/62712.

254. *The hidden water resource use behind meat and dairy.* Hoekstra, Arjen Y. 2, 2 April 2012, Animal Frontiers, Vol. 2, pp. 3-8.

255. C. R, Dahlan, J. B. Lamb. *Livestock Water Requirements and Water Quality.* s.l. : North Dakota State UNiversity Extension, 2009.

256. *Invited review: Freedom from thirst-Do dairy cows and calves have sufficient access to drinking water?* Jensen MB, Vestergaard M. 11, 11 August 2021, Journal of Dairy Science, Vol. 104, pp. 11368-11385.

257. Nations, Food and Agriculture Organization of the United. *Greenhouse Gas Emissions from the Dairy Sector A Life Cycle Assessment.* s.l. : FAO.

258. *Livestock, Deforestation, and Policy Making: Intensification of Cattle Production Systems in Central America Revisited.* Charles F. Nicholson, Robert W. Blake, David R. Lee. 3, March 1995, Journal of Dairy Science, Vol. 78, pp. 719-734.

259. *The grey water footprint of milk due to nitrate leaching from dairy farms in Canterbury, New Zealand.* M. K. Joy, D. A. Rankin, L. Wöhler, P. Boyce, A. Canning, K. J. Foote, P. M. McNie. 2, 2022, Australian Journal of Environmental Management, Vol. 29, pp. 177-199.

260. Ireland, Environment. OEP: Agri-industry leading driver of biodiversity decline. *Environment Ireland.* [Online] 22 November 2024. https://www.environmentireland.ie/oep-agri-industry-leading-driver-of-biodiversity-decline/.

261. *An exploration of biodiversity limits to grazing ruminant milk and meat production.* Kajsa Resare Sahlin, Line J. Gordon, Regina Lindborg, Johannes Piipponen, Pierre Van Rysselberge, Julia Rouet-Leduc, Elin Röös. 25 July 2024, Nat Sustain, Vol. 7, pp. 1160-1170.

262. Initiative, Dairyland. Manure Management. *Dairyland Initiative .* [Online] https://thedairylandinitiative.vetmed.wisc.edu/home/housing-module/adult-cow-housing/manure-management/.

263. *Cheese Yield.* Banks, J. M. s.l. : Woodhead Publishing, 2007, Cheese Problems Solved, pp. 100-114.

264. *Recent advances in why processing and valorisation: Technological and environmental perspectives.* D. Buchananan, W. Martindale, E.

Romeih, E. Hebishy. 4, 2023, International Journal of Dairy Technology, Vol. 76, pp. 659-670.

265. *Whey Utilization: Sustainable Uses and Environmental Approach.* Elizabeta Zandona, Marijana Blažić, Anet Režek Jambrak. 2, June 2021, Food Technol Biotechnol, Vol. 59, pp. 147-161.

266. Education, Department of. School food standards: Practical guide. *GOV.UK.* [Online] [Cited: 13 February 2025.] https://www.gov.uk/government/publications/school-food-standards-resources-for-schools/school-food-standards-practical-guide.

267. Knips, V. Developing Countries and the Global Dairy Sector: Part 1 - Global overview (PPLPI Working Paper No. 23768. s.l. : Food and Agriculture Organization of the United Nations, 2005.

268. Dairy consumption and risks of total and site-specific cancers in Chinese adults: an 11-year prospective study of 0.5 million people. Kakkoura, M.G., Du, H., Guo, Y. et al. 10 March 2022, BMC Medicine, Vol. 20.

269. High glycemic load diet, milk and ice cream consumption are related to acne vulgaris in Malaysian young adults: a case control study. Ismail, N.H., Manaf, Z.A. & Azizan, N.Z. 9 August 2012, BMC Dermatology, Vol. 12.

270. *Systematic review of the epidemiology of acne vulgaris.* Heng, Anna Hwee Sing, and Fook Tim Chew. 1, 1 April 2020, Scientific Reports, Vol. 10.

271. *Dietary modifications in atopic dermatitis: patient-reported outcomes.* Nosrati, Adi et al. 6, 1 September 2018, The Journal of Dermatology Treatment, Vol. 28, pp. 523-538.

272. Plasma Concentrations of Trimethylamine-N-oxide Are Directly Associated with Dairy Food Consumption and Low-Grade Inflammation in a German Adult Population. Rohrmann Sabine, Linseisen Jakob, Allenspach Martina, von Eckardstein Arnold, Müller Daniel. 2, February 2016, The Journal of Nutrition, Vol. 146, pp. 283-289.

273. A review of dairy food intake for improving health for black women in the US during pregnancy, fetal development, and lactation. 2, April 2024, Journal of the National Medical Association, Vol. 116, pp. 219-227.

274. *The Myth of Increased Lactose Intolerance in African-Americans.* Byers, K. G. & Savaiano, D. A. (sup6), December 2005, Byers, K. G., & Savaiano, D. A. (2005). The Myth of Increased Lactose Intolerance in AJournal of the American College of Nutrition, Vol. 24, pp. 569S-573S.

275. *A bacteriological method for determining manural pollution of milk.* Weinzirl, John, and Milton V. Veldee. 1915, Amcerican Journal of Public Health, Vol. 5.9, pp. 862-866.

276. A note on the Weinzirl Anaerobic Spore Test for Determining Manurial Pollution of Milk. Hudson, James E. et al. 4, Journal of Dairy Science, Vol. 5, pp. 377-382.

277. *Plant-Based Formulas and Liquid Feedings for Infats and Toddlers.* Vandenplas Y, De Mulder N, De Greef E, Huysentruyt K. 11, November 2021, Nutrients, Vol. 13.

278. *Making Fashion Sustainable: Waste and Collective Responsibility.* Moorhouse, Debbie. 1, 24 July 2020, One Earth, Vol. 3, pp. 17-19.

279. *A new textiles economy: Redesigning fashion's future.* Foundation, Ellen MacArthur. s.l. : Ellen MacArthur Foundation, 2017.

280. Commission, International Sericultural. Statistics. *International Sericultural Commission.* [Online] https://inserco.org/en/statistics.

281. America, Council of Fashion Desingers of. Fiber Guide Silk. *CFDA Materials Hub.* [Online] 13 October 2021. https://cfda.com/resources/materials-hub/article/fiber-guide-silk/.

282. Fritz, Anne and Cant. *Consumer Textiles.* s.l. : Oxford University Press Australia, 1986. ISBN 0-19-554647-4.

283. Editor, Fabric. The Environmental Impact of Silk Fabric Production. *Fabric Material Guide.* [Online] https://fabricmaterialguide.com/the-environmental-impact-of-silk-fabric-production/?utm.

284. Australia, Meat and Livestock. *Australia's sheepmeat industry.* 2022. p. 1.

285. Feed efficiency for meat and wool production by Merino and F1 Dohne x Merino lambs fed pelleted diets of different nutritive value. Van Beem, D., Wellington, D., Paganoni, B. L., Vercoe, P., & Milton, J. 7, 2008, Animal Production Science, Vol. 48, pp. 879-884.

286. Jackson, Tim. Australian lamb and mutton dominate global markets. *Meat and Livestock Australia.* [Online] 5 April 2024. https://www.mla.com.au/news-and-events/industry-news/australian-lamb-and-mutton-dominate-global-markets/.

287. FAO. *Livestock's Long Shadow - environmental issues and options.* Rome : Food and Agriculture Organization of the United Nations, 2006.

288. Food in the Anthropocene: the EAT-Lancet Commission on healthy diets from sustainable food systems. Willett W, Rockström J, Loken B, et al. February 2019, LAncet, Vol. 393, pp. 477-492.

289. Sustainable Strategies for Greenhouse Gas Emission Reduction in Small Ruminants Farming. Giamouri, E., et al. 5, 2023, Sustainability, Vol. 15, p. 4118.

290. Executive, Health and Safety. Agricultural Information SheetNo. 41, Sheep Dipping Advice for Farmers and Others Involved in Dipping Sheep. *Health and Safety Executive.* [Online] 2013. http://www.hse.gov.uk.pubns/ais41.pdf.

291. *Organophosphate Poisoning: Review of Prognosis and Management.* Zoofaghari, Shafeajafar et al. 23 September 2024, Advanced biomedical research, Vol. 13.

292. DEFRA. Sheep dip: groundwater protection code. February 2023.

293. Bernie Thomas, Matt Fishwick, James Joyce and Anton van Santen. Table 21 "A Carbon Footprint for UK Clothing and Opportunities for Savings".

294. The water footprint of cotton consumption: An assessment of the impact of worldwide consumption of cotton products on the water resources in the cotton producing countries. Chapagain, A., Hoekstra, A. Y., Savenije, H. H. G., & Gautam, R. 1, 2006, Ecological Economics, Vol. 60, pp. 186-203.

295. *The green, blue and grey water footprint of crops and derived crop.* Mekonnenm M. M., Hoekstra, A. Y. 2011, Hydrology and Earth System Sciences.

296. Flax, European. Life Cycle Assessment of European Flax Scutched Fibre. 2023.

297. Exchange, The Textile. Materials Dashboard. s.l. : Textile Exchange.

298. Nguyen, Quynh. How Sustainable Are Linen Fabrics? A Life-Cycle Analysis. *Impactful Ninja.* [Online] https://impactful.ninja/how-sustainable-are-linen-fabrics/.

299. Angela Bernhardt, Dr. Jack Caravanos, Richard Fuller OAM, Stephen Leahy and Ami Pradhan. *Pollution Knows No Borders.* s.l. : Blacksmith Institute, 2019.

300. *Life Cycle Assessment of Cow Tanned Leather Products.* M Ulya, A L Arifuddin and K Hidayat. 25 November 2020, IOP Conference Series: Earth and Environmental Science, Vol. 757.

301. Kering. Kering Environmental Profit & Loss Report 2021 Group Results. 2022.

302. —. Kering Environmental Profit & Loss Report 2018. 2019.

303. Index, Higgs. Higg Index.

304. *Environmental impact assessment of man-made cellulose fibres.* Shen, Li et al. 2, December 2010, Resources, Conservation and Recycling, Vol. 55, pp. 260-274.

305. *EDIPTEX - Environmental assessment of textiles.* Laursen, Søren & Hansen, John & Knudsen, Hans & Wenzel, Henrik & Larsen, Henrik & Kristensen, Frans. January 2007.

306. Fact, Carbon. Carbon Fact Resources.

307. Platform, European Circular Economy Stakeholder. Consultation on the Product Environmental Footprint Category Rules for apparel and footwear now open!

308. Renton, Alex. Bangladesh's toxic tanneries turning a profit at an intolerable human price. *The Guardian.* [Online] 2012. [Cited: 13 March 2025.] https://www.theguardian.com/global-development/2012/dec/13/bangladesh-toxic-tanneries-intolerable-human-price.

309. *Assessment of selected metals (chromium, lead and cadmium) in the hair of tannery workers at Hemayetpur, Bangladesh.* Nurjahan Akter

Shimo, Md. Abdus Salam, Maksuda Parvin, Md. Zakir Sultan. June 2023, Journal of Trace Elements and Minerals, Vol. 4.

310. *Mitigating Tannery Pollution in Sub-Saharan Africa and South Asia.* Hira, A., Pacini, H., Attafuah-Wadee, K., Sikander, M., Oruko, R., & Dinan, A. 3, 2022, Journal of Developing Societies, Vol. 38, pp. 360-383.

311. Alliance, Fur Free. Covid 19 on Mink Farms. *Fur Free Alliance.* [Online] https://www.furfreealliance.com/covid-19-on-mink-farms/.

312. Pickett, Heather and Harris, Stephen. The case against fur factory farming: A scientific review of animal welfare standards and 'WelFur'. s.l. : Euro Group for Animals, 2023.

313. Needle Structures and Epiphytic Microflora of Scots Pine (Pinus Sylvestris L.) under Heavy Ammonia Deposition from Fur Farming. Bäck, J., Turunen, M., Ferm, A. et al. 1997, Water, Air, & Soil Pollution, Vol. 100, pp. 119-132.

314. Buchholz, Katharina. Global Fur Production in Drastic Decline. *Statista.* [Online] 3 February 2025. [Cited: 27 February 2025.] https://www.statista.com/chart/33876/supply-volumes-of-mink-fox-pelts-on-the-world-market/.

315. PETA. Wool, Fur, and Leather: Hazardous to the Environment. *PETA.* [Online] https://www.peta.org/issues/animals-used-for-clothing-factsheets/wool-fur-leather-hazardous-environment/.

316. Alliance, Fur Free. Toxic Fur. [Online] https://www.furfreealliance.com/toxic-fur/.

317. ECOPEL. The Truth About Faux Fur and the Sustainability. [Online] [Cited: 27 February 2025.] https://www.ecopel.com/article-03---the-truth-about-faux-fur---sustainability.html?.

318. Brucculieri, Julia. Faux Fur Is Made Of Plastic, And It's Not Helping The Environment. *HuffPost.* [Online] 28 February 2019. [Cited: 27 February 2025.] https://www.huffingtonpost.co.uk/entry/faux-fur-vs-real-fur_n_5bc0b3c3e4b0bd9ed5599f76.

319. Brownlee, Charlotte. Meet GACHA: a new biodegradable faux fur. *Collective Fashion Justice.* [Online] [Cited: 27 February 2025.]

https://www.collectivefashionjustice.org/articles/gacha-biodegradable-faux-fur.

320. NASA. 11.0 Spacesuits. *NASA Explore.* [Online] 13 February 2025. [Cited: 27 February 2025.] https://www.nasa.gov/reference/11-0-spacesuits-vol-2/.

321. Blaustein-Rejto, Dan. Livestock Don't Contribute 14.5% of Global Greenhouse Gas Emissions. *The Breaakthrough Institute.* [Online] 20 MArch 2023. https://thebreakthrough.org/issues/food-agriculture-environment/livestock-dont-contribute-14-5-of-global-greenhouse-gas-emissions.

322. Ritchie, H., Roser, M., & Rosado, P. CO_2 and Greenhouse Gas Emissions . *Our World in Data.* [Online] 11 May 2020. https://ourworldindata.org/co2-and-other-greenhouse-gas-emissions.

323. IPCC. Climate Change 2022: Mitigation of Climate Change. Contribution of Working Group III to the Sixth Assessment Report of the Intergovernmental Panel on CLimate Change. *IPCC.* [Online] https://www.ipcc.ch/report/ar6/wg3/.

324. (n.d.), World Resources Institute. CLimate Watch: Historical GHG Emissions. *Climate Watch.* [Online] https://www.climatewatchdata.org/ghg-emissions.

325. Agency, International Energy. Greenhouse Gas Emissions from Energy: Overview. *IEA.* [Online] 2021. https://www.iea.org/reports/greenhouse-gas-emissions-from-energy-overview.

326. Nations, Food and Agriculture Organization of the United. Tackling Climate Change Through Livestock: A Global Assessment of Emissions and Mitigation Opportunities. *FAO.* [Online] 2013. http://www.fao.org/3/i3437e/i3437e.pdf.

327. Ritchie, Hannah. Sector by sector: where do global greenhouse gas emissions come from? *OurWorldinData.org.* [Online] 2020. https://ourworldindata.org/ghg-emissions-by-sector.

328. Crippa, Monica, et al. EDGAR 2024 Greenhouse Gas Emissions. *European Commission, Joint Research Centre (JRC) [Dataset].* [Online] 2024. http://data.europa.eu/89h/88c4dde4-05e0-40cd-a5b9-19d536f1791a.

329. Washington, DC: World Resources Institute. Climate Watch Historical GHG Emissions. *World Resources Institute.* [Online] 2022. https://www.climatewatchdata.org/ghg-emissions.

330. To what extent can zero tillage lead to a reduction in greenhouse gas emissions from temperate soils? Mangalassery, S., Sjögersten, S., Sparkes, D. et al. 4 April 2014, Scientific Reports, Vol. 4.

331. India, The Times of. Ghazipur landfill collapse. *The Times of India.* [Online] 1 September 2017. https://timesofindia.indiatimes.com/calamities/ghazipur-landfill-collapse/articleshow/60325175.cms:contentReference{index=0}.

332. Lazarus, N. Ghazipur landfill: The 70-acre 'garbage mountain of Dehli'-where nearby residents are being 'slowly poisoned'. *Sky News.* [Online] 8 March 2025. [Cited: 15 March 2025.] https://news.sky.com/story/ghazipur-landfill-the-70-acre-garbage-mountain-of-delhi-where-nearby-residents-are-being-slowly-poisoned-13323429.

333. (nd.), Kayrros. Methane Watch. *Kayrros.* [Online] [Cited: 15 March 2025.] https://methanewatch.kayrros.com/map.

334. WWF-UK. Keeping it cool: How the UK can end its contribution to climate change. *WWF-UK.* [Online] 2018. [Cited: 25 March 2025.] https://www.wwf.org.uk/sites/default/files/2018-11/NetZeroReportART.pdf.

335. (IPCC), Intergovernmental Panel on Climate Change. Climate Change 2022: Mitigation of Climate Change. Contribution of Working Group III to the Sixth Assessment Report of the Intergovernmental Panel on Climate Change. *IPCC.* [Online] 2022. [Cited: 21 March 2025.] https://www.ipcc.ch/report/ar6/wg3/.

336. Drawdown, Project. Climate Solution for a New Decade. *Project Drawdown.* [Online] 2020. [Cited: 21 March 2025.] https://drawdown.org/drawdown-review.

337. (IEA), International Energy Agency. Net Zero by 2050: A Roadmap for the Global Energy Sector. *IEA.* [Online] 2021. [Cited: 21 March 2025.] https://www.iea.org/reports/net-zero-by-2050.

338. Company, McKinsey &. The Net-Zero Transition: What It Would Cost, What It Could Bring. *McKinsey.* [Online] 2022. [Cited: 21 March 2025.] https://www.mckinsey.com/business-functions/sustainability/our-insights/the-net-zero-transition-what-it-would-cost-what-it-could-bring.

339. (FAO), Food and Agriculture Organization of the United Nations. Food Wastage Footprint: Impacts on Natural Resources. *FAO.* [Online] 2013. [Cited: 21 March 2025.] https://www.fao.org/3/i3347e/i3347e.pdf.

340. *Natural climate solutions.* Griscom, B. W., Adams, J., Ellis, P. W., Houghton, R. A., Lomax, G., Miteva, D. A., ... & Fargione, J, 114(44), 0. 44, 16 October 2017, Proceedings of the National Academy of Sciences, Vol. 114, pp. 11645–11650.

341. London, WWF-UK & Imperial College. Behaviour Change, Public Engagement and Net Zero. *WWF-UK.* [Online] 2020. [Cited: 21 March 2025.] https://www.wwf.org.uk/sites/default/files/2020-03/WWF%20-%20Net%20Zero%20-%20Behaviour%20Change%20-%20Final.pdf.

342. Stress, anxiety, and depression in times of COVID-19: Gender, individual quarantine, pandemic duration and employment. Levy, I. (2020). ., 8, 999795. s.l. : Frontiers in Public Health, 2020.

343. Depression as a psychosocial consequence of occupational injury in the US working population: Findings from the Medical Expenditure Panel Survey. Kim, J. 2013, BMC Public Health, Vol. 13, p. 303.

344. Prevalence of serious psychological distress among slaughterhouse workers at a United States beef packing plant. Leibler, J. H., Janulewicz, P. A., & Perry, M. J. 1, 2017, Work, Vol. 51, pp. 105–109.

345. *Suicide Rates by Industry and Occupation — National Vital Statistics System, United States, 2021.* Sussell A, Peterson C, Li J, Miniño A, Scott KA, Stone DM. 15 December 2023, Morbidity and Mortality Weekly Report , Vol. 72, pp. 1346–1350.

346. *Rural male suicide in Australia.* M, Alston. Feb 2012, Social science & medicine, Vol. 74, pp. 515–522.

347. Understanding the factors contributing to farmer suicide: a meta-synthesis of qualitative research. Purc-Stephenson, R., Doctor, J., & Keehn, J. E. August 2023, Rural and remote health, Vol. 23.

348. Farmer welfare and animal welfare- Exploring the relationship between farmer's occupational well-being and stress, farm expansion and animal welfare. Hansen, B. G., & Østerås, O. 1 October 2019, Preventive Veterinary Medicine, Vol. 170.

349. IPCC. Intergovernmental Panel on Climate Change (IPCC). (2018). Global Warming of 1.5°C: An IPCC Special Report on the impacts of global warming of 1.5°C above pre-industrial levels and related global greenhouse gas emission pathways. . s.l. : The World Counts, 2018.

350. *Six degrees could change the world.* National Geographic Channel, 2008.

351. Peter Aldhous, Stephanie M. Lee, Zahra Harji. The Texas Winter Storm And Power Outages Killed Hundreds More People Than The State Says. *Buzzfeed News.* [Online] 26 March 2021. https://www.buzzfeednews.com/article/peteraldhous/texas-winter-storm-power-outage-death-toll.

352. Roy P. E. Yanong, Ruth Francis-Floyd, Barbara D. Petty. Infectious Diseases in Aquaculture. *MSD Manual Veterinary Manual.* [Online] October 2021. https://www.msdvetmanual.com/exotic-and-laboratory-animals/aquaculture/infectious-diseases-in-aquaculture.

353. Times, Tasmanian. Current Affairs Communities Ramp-up Campaign Against Salmon Farming Expansion. *TasmanianTimes.com.* [Online] 6 December 2024. https://tasmaniantimes.com/2024/12/communities-ramp-up-campaign-against-salmon-farming-expansion/.

354. *Isolation, characterization and comparison of bacteria from swine faeces and manure storage pits.* Michael A Cotta, Terence R Whitehead, Rhonda L Zeltwanger. s.l. : Environ Microbio, 5 September 2003, pp. 737-745.

355. Hannah Ritchie, Pablo Rosado, Max Roser. Environmental Impacts of Food Production. *Our World in Data.* [Online] 2022. https://ourworldindata.org/environmental-impacts-of-food.

356. Inner Mongolia Yili Industrial Group Co., Ltd. Department of Economic and Social Affairs. *United Nation Department of Economic and Social Affairs.* [Online] https://sdgs.un.org/partnerships/low-water-footprint-initiative-lwfi-dairy-industry.

357. Trust, Oxford Health NHS Foundation. Cholesterol. *Oxford Health NHS Foundation Trust.* [Online] 2014. https://www.oxfordhealth.nhs.uk/wp-content/uploads/2014/08/OP-026.14-Cholesteral.pdf.

358. Peterson, LeLyn. How Much Does a Leather Jacket Weigh. *Leather Insights.* [Online] https://leatherinsights.com/how-much-does-a-leather-jacket-weigh/.

359. Eutrophication and recovery in experimental lakes: implications for lake management. W., Schindler D. 24 May 1974, Science, Vol. 184, pp. 897–899.

About the author

Montoya Whitehead is a passionate advocate for animals, the planet, and human health. Drawing on extensive research and a commitment to justice, she writes to uncover the hidden costs of animal agriculture. She also seeks to highlight the transformative potential of a plant-based world. Her goal is to empower readers with knowledge and inspire compassionate, sustainable change with a respect for all life.

Through *The Young Vegan* (www.theyoungvegan.co.uk), Montoya shares resources, articles, and insights to help people transition toward compassionate, sustainable living. Her work encourages readers to learn, grow, and feel connected to all life on the planet. Inspiring them to live in a way that is compassionate and sustainable.

About the artist

Francisco Atencio is a freelance illustrator and visual artist from Argentina. His paintings focus on issues that are vital to him: animal rights, humanity and its impact on our planet, and the beauty of landscapes. His illustrations have appeared in several books dedicated to environmental protection and the benefits of a plant-based lifestyle.

www.ingramcontent.com/pod-product-compliance
Lightning Source LLC
Chambersburg PA
CBHW051523020426
42333CB00016B/1749